AF874687

The Urban Book Series

Aims and Scope

The Urban Book Series is a resource for urban studies and geography research worldwide. It provides a unique and innovative resource for the latest developments in the field, nurturing a comprehensive and encompassing publication venue for urban studies, urban geography, planning and regional development.

The series publishes peer-reviewed volumes related to urbanization, sustainability, urban environments, sustainable urbanism, governance, globalization, urban and sustainable development, spatial and area studies, urban management, urban infrastructure, urban dynamics, green cities and urban landscapes. It also invites research which documents urbanization processes and urban dynamics on a national, regional and local level, welcoming case studies, as well as comparative and applied research.

The series will appeal to urbanists, geographers, planners, engineers, architects, policy makers, and to all of those interested in a wide-ranging overview of contemporary urban studies and innovations in the field. It accepts monographs, edited volumes and textbooks.

More information about this series at http://www.springer.com/series/14773

Carlos Nunes Silva · Ján Buček
Editors

Local Government and Urban Governance in Europe

Springer

Editors
Carlos Nunes Silva
Institute of Geography and Spatial Planning
Universidade de Lisboa
Lisbon
Portugal

Ján Buček
Department of Human Geography
and Demography
Comenius University
Bratislava
Slovakia

ISSN 2365-757X ISSN 2365-7588 (electronic)
The Urban Book Series
ISBN 978-3-319-43978-5 ISBN 978-3-319-43979-2 (eBook)
DOI 10.1007/978-3-319-43979-2

Library of Congress Control Number: 2016948116

© Springer International Publishing Switzerland 2017
This work is subject to copyright. All rights are reserved by the Publisher, whether the whole or part of the material is concerned, specifically the rights of translation, reprinting, reuse of illustrations, recitation, broadcasting, reproduction on microfilms or in any other physical way, and transmission or information storage and retrieval, electronic adaptation, computer software, or by similar or dissimilar methodology now known or hereafter developed.
The use of general descriptive names, registered names, trademarks, service marks, etc. in this publication does not imply, even in the absence of a specific statement, that such names are exempt from the relevant protective laws and regulations and therefore free for general use.
The publisher, the authors and the editors are safe to assume that the advice and information in this book are believed to be true and accurate at the date of publication. Neither the publisher nor the authors or the editors give a warranty, express or implied, with respect to the material contained herein or for any errors or omissions that may have been made.

Printed on acid-free paper

This Springer imprint is published by Springer Nature
The registered company is Springer International Publishing AG Switzerland

Contents

List of Figures

List of Tables

Chapter 1
Introduction

Carlos Nunes Silva and Ján Buček

Europe has in 2015 a population of nearly 743 million inhabitants, of which around 74 % live in urban areas, according to the 2014 United Nations World Urbanization Prospects. While the total population in the continent is projected to decrease in the next decades, expected to fall to 709 million in 2050, the same number of inhabitants the continent had in 1985, it will continue to urbanize, and as a consequence, the proportion of urban population is expected to reach 82 % in 2050.[1] In 2015, Europe has 3 megacities, urban agglomerations with more than 10 million inhabitants, 2 urban agglomerations between 5 and 10 million inhabitants, and 48 between 1 and 5 million inhabitants, a number expected to increase up to 55 in 2050, according to the same United Nations source. These urban patterns and trends differ from country to country and among regions within the European continent and confront local governments in Europe, in different ways, with new challenges and opportunities in the governance of cities.

The book aims to discuss innovative reforms and practices developed and experimented in critical areas of urban governance in the last decades, in part associated with the impact of the European Union policies on its member states, or associated with the transition to democracy, which in some southern European countries occurred in the 1970s and in central, eastern and southeast Europe in the

[1]The proportion of urban population in the world is 54 % in 2015 and is projected to reach 66.4 % in 2050 (UN-World Urbanization Prospects. The 2014 Revision. New York: United Nations). These figures used here refer to the aggregated data considered as Europe in the United Nations World Urbanization Prospects.

C. Nunes Silva (✉)
Institute of Geography and Spatial Planning, Universidade de Lisboa,
Lisbon, Portugal
e-mail: cs@campus.ul.pt

J. Buček
Department of Human Geography and Demography,
Comenius University, Bratislava, Slovakia

© Springer International Publishing Switzerland 2017

C. Nunes Silva and J. Buček (eds.), *Local Government and Urban Governance in Europe*, The Urban Book Series, DOI 10.1007/978-3-319-43979-2_1

early 1990s, in this case as part of the political process that followed the fall of the socialist political bloc in eastern and central Europe. The book deals primarily with the concepts of local government and urban governance. The notion of local government is used in the book to refer subnational tiers of government, whose nature varies from country to country, a variation that reflects the idiosyncrasy of the national political processes and the prevailing public administration cultures as well. We distinguish basically two models of local government, associated one with deconcentration and the other with decentralization. The first corresponds to the local administration of the state, which continues to be an important stakeholder in urban governance in these countries; the second refers the true form of local government, which we term local self-government, with political, administrative, and financial autonomy and with directly elected boards, among other characteristics. Across the book, local self-government is referred by a variety of terms that better reflect the specific national context of the case examined in each chapter and the professional background of each author. All these different terms refer the notion of local self-government, and we use this and local government as synonymous throughout the book. The other concept is urban governance. By this, we refer the mode of governing cities that is emerging as a result of the move from the traditional hierarchical public administration to a mode of governing based increasingly on a network of public and nonpublic entities, of which the state, both central government and the local administration of the state, as well as local self-government, are still the key stakeholders. Again, the exact configuration of these institutional architectures varies from country to country, and this is reflected somehow in the variety of terms used across the book. Important to retain is the fact that in none of the chapters in the book is the governing of urban areas seen as being carried out solely by traditional hierarchical public administration.

The methodology used varies between chapters but the book uses in most of its twelve chapters a qualitative case study approach, relying on different types of data, collected in official policy reports, plans, background policy documents, in national and international legislation, in statistical publications, in previous academic research, or through interviews, public opinion surveys, external observation, or in participatory workshops. If it is true that the small number of cases examined may be seen as a limitation of the book, affecting its external validity, it is also true that the qualitative approach applied in these case studies allowed the identification of new relationships and the exploration of new perspectives, and at the same time a deeper knowledge of each individual case studied.

The book has 12 chapters, including this Introduction, and is organized into three parts. The first part—"Crisis, reform and innovation in local government"—has four chapters focused on the development of local self-government in Europe. The first three chapters illustrate well two processes responsible for the most important changes experienced by local self-government in Europe in the last half century, the transition to democracy in southern Europe in the mid-1970s and in central and eastern Europe in the early 1990s, as well as the impacts of the 2008 global economic and financial crisis in the restructuration of local self-government systems in a large number of European countries. In Chap. 2, "Political and Administrative

Decentralization in Portugal: Four Decades of Democratic Local Self-Government," the book examines the main changes in the process of political and administrative decentralization and in the development of local self-government in Portugal since the transition to democracy in 1974, seen within the wider institutional changes that affected European countries, in particular since Portugal joined the European Community in 1986. The chapter focuses, in particular, the period during which Portugal was under the Memorandum of Understanding on specific economic policy conditionality, and the proposals of the XXI Government (2015–2019) for the reform of local government, which are expected to revert part of the policy options taken during the recent economic adjustment program (2011–2014/2015). Ján Buček in Chap. 3—"Why and When Countries Implement Local Public Administration Reforms: A Long-Term View of Reform Dynamics in Slovakia 1990–2015"—explores the conditions behind reforms and adaptations experienced by local self-government in Slovakia after the transition to democracy in 1989 and argues that reforms in public administration, including in local government, required a highly complex set of conditions to be successful. The chapter on Portugal illustrates the main institutional changes that countries in southern Europe experienced since the mid-1970s, as part of the transition from authoritarian political regimes to democracy, as well as in reaction to the 2008 global economic and financial crisis. The chapter on Slovakia does the same for the transition to democracy experienced by countries in central and eastern Europe, constituting these two transitions processes the most important political shifts in the last half century in Europe.

The next chapter "Local Self-Government in Hungary: The Impact of Crisis," by Adrián Fábián, discusses the impact of the 2008 economic and financial crisis on local self-government in Hungary and argues that local government was already affected by important financial constraints, which were not originated by the 2008 crisis, as this only increased the problems that already existed. The chapter examines also the reforms implemented in response to these problems, and the results achieved until 2015. The last chapter in this first part deals with specific institutional changes in local and regional government in southern Europe. Chapter 5, "Local Government Innovation in Italy: Impact on Urban and Regional Planning with a focus on the Milanese context," written by Renzo Riboldazzi, examines the impacts of the 2014 reform of local government in Italy and gives particular attention to the regional governance and planning innovations, taking Milan as a case study for an in-depth analysis of this recent local government reform process.

The second part of the book—"EU Policies, the Urban Agenda and Local Governance"—has five chapters that deal each with a specific issue within the European Union (EU) urban policies or urban agenda. In the first of these four chapters, "Urban dimension of the European policies and the new EU Urban Agenda in Croatia," Dana Dobrić examines policies and standards for good urban governance adopted by the European Union and by the Council of Europe, giving special attention to local development policies and to the role given to cities in their implementation and ends with an analysis of the Croatian case as an example of the challenges and opportunities confronting cities in Europe within the emerging Europe 2020 governance framework. The following chapter, "The relationship between the European Commission and Local Governments through European Urban Initiatives: constraints and solidarities," by Oussama Kharchi, puts into perspective the European urban initiatives, in order to understand the changes in the relations between the European Commission and cities in Europe and how it impacted on urban governance. In Chap. 8, "Mega-events and new patterns of cooperation: the European Capitals of Culture," Ágnes Németh examines the opportunities for social capital mobilization in European cities in the context of the European Capital of Culture program, launched 30 years ago, based on a comparison of two cases, Pécs, in Hungary, and Turku, in Finland. The author explores how this EU program facilitated and changed inter-organizational cooperation and networking among cultural and creative producers in these cities, pointing out the importance of knowledge and support networks in the field of creative industries and the role of these large-scale cultural events in the formation and on the sustainability of such networks, so important for urban governance. Anna Trono, Maria Chiara Zerbi, and Valentina Castronuovo examine, in Chap. 9, "Urban Regeneration and Local Governance in Italy. Three emblematic cases," the role of urban renewal policies, in particular those associated with the EU Urban Initiative program, in the regeneration of urban areas, based on three case studies: Lecce and Taranto, in the south of Italy, and Milan, in the north. As the authors show, there are sharp differences in the results achieved so far due to the differences in management capacity of local administrations and public and private partners as well. Finally, in Chap. 10, "Involving citizens in the reuse and regeneration of urban peripheral spaces," Beniamino Murgante and co-authors examine an experience of participatory reactivation of abandoned or underused spaces and buildings in a neighborhood in Potenza, Basilicata, Italy. This participatory planning experience shows how relatively simple and low-cost urban regeneration interventions carried out with citizens engagement can have better results than highly expensive and complex traditional urban regeneration programs.

The third part of the book—"Citizen Participation in Local Self-Government"—comprises two chapters and explores a key component in urban governance: citizen participation in local government decision-making. The two chapters deal with citizen participation in local elections. Lilita Seimuskane, Inga Vilka, and Edgars Brekis in Chap. 11 "Assessment of socio-economic status relevance for Latvian electoral participation" explore the factors that affect the level of electoral participation in Latvia, the choice to vote and the motivation to participate in elections pointing for the existence of a correlation between socio-economic status and participation in local government elections in Latvia, concluding that while socio-economic factors are more clearly correlated with participation in the election to the European Parliament, in the local elections the contextual factor, specific of each municipality, has an important influence in the electoral outcome. The authors compare the electoral participation in the local elections of 2013 with the elections to the European Parliament in 2014. The following chapter "Making informed citizens in local direct democracy. What part does their government perform?," written by Werner Pleschberger, discusses the issue of a voter' competence in direct democracy processes at the level of local government decision-making processes, based on an analysis and comparison of three cases: the Austrian cities of Vienna and Bregenz and the city of Los Angeles in the USA. Included in the discussion are the practical ways of how local government can inform the voter on the fiscal implications of specific measures they would have to decide on the ballot, concluding that Los Angeles has a high developed approach of voter information and Vienna the poorest regulation and practice of voter information, with the city of Bregenz representing a moderate example.

Changes in the local self-government system and reforms in local self-government policies, as those examined in these essays show, require the combination of numerous factors and highly complex incentives. The evidence collected suggests that the success of local government reforms is dependent on national conditions, namely the administrative culture, the pattern of central–local relationships, the political and economic conditions, the existence of a balanced public finance system, among other factors, and international influences as well, namely through the European Union and Council of Europe. Besides the influence of these two layers, local government reform and policy practices are highly dependent on and influenced by local contexts, namely by the differences in management capacity of local government. The evidence provided in these essays shows that the road for reform and the outcomes tend to differ substantially from country to country, or among cities, since national and local contexts are determinant factors in the functioning of local government, as the cases of Portugal, Italy, Croatia, Hungary, or Slovakia illustrate. European integration and the implementation of EU policies in different sectors have been responsible for the implementation of structural reforms in the local government systems of member states, a process that is now starting to be fully experienced in the new EU members, as the case of Croatia illustrates. The recognition of cities as a relevant place for the European Union intervention reshaped the relationships between the EU and local government in each member state, as well as the institutional models and practices of urban governance, as some of these chapters show. The importance of local self-government is so critical that failure in its reform risks to have serious negative consequences for the community, as the analysis of the recent reform of local government in Italy, or those in Portugal, Slovakia, and Hungary point out.

In sum, by providing this informed and updated view of recent institutional reforms and the key challenges confronting the governance of cities in Europe, the book will be of interest for students and researchers in the broad field of urban studies, in particular for all those working on urban governance in European countries and for policy makers in central and in local government as well.

Part I
Crisis, Reform and Innovation in Local Government

Chapter 2
Political and Administrative Decentralization in Portugal: Four Decades of Democratic Local Government

Carlos Nunes Silva

Abstract The chapter examines and discusses the process of political and administrative decentralization in Portugal seen as the basic institutional framework for urban governance. It focuses, in particular, the period during which Portugal was under the Memorandum of Understanding on specific economic policy conditionality (MoU) and the proposals of the XXI Government (2015–2019) for the reform of local government, which are expected to revert part of the policy options taken during the recent economic adjustment programme (2011–2014/2015). The research on which the chapter is based uses a qualitative case study approach. The chapter shows that the transition from the authoritarian period (1926–1974)—the military dictatorship (1926–1932) and the authoritarian political regime of the Estado Novo (1933–1974)—to the II Republic, in 1974, marks a shift in the nature and in the modus operandi of sub-national regional and local self-government in Portugal. It also shows that these institutional changes and social progress, the result of decentralization from the state to sub-national tiers of government, were affected by the austerity policy implemented by the XIX Government (2011–2015) in the context of the assistance program. This counterrevolution in the local government system, produced in the name of an austerity policy imposed by international institutions in the context of the MoU, is expected to be reverted, at least partially, by the XXI Government, considering the electoral manifestos of the three political parties that support it and the government program approved in Parliament.

Keywords Decentralization · Local autonomy · Local self-government · Regional and metropolitan government · Inter-municipal cooperation · Urban governance · Portugal

C. Nunes Silva (✉)
Institute of Geography and Spatial Planning,
Universidade de Lisboa, Lisbon, Portugal
e-mail: cs@campus.ul.pt

© Springer International Publishing Switzerland 2017

C. Nunes Silva and J. Buček (eds.), *Local Government and Urban Governance in Europe*, The Urban Book Series, DOI 10.1007/978-3-319-43979-2_2

2.1 Introduction

Local government in Portugal went through important changes in its nature and structure in the last four decades, starting with the transition to democracy in the 1970s and continuing through successive political reforms since then. The transition from the authoritarian period (1926–1974)—the military dictatorship (1926–1932) and the authoritarian political regime of the Estado Novo (1933–1974)—to the II Republic, in 1974, marks a rupture in the nature and in the modus operandi of sub-national regional and local self-government in Portugal. During the period of the authoritarian regime (1926–1974), sub-national tiers of government were strictly dependent and controlled by central government. There were no direct and free elections for the local boards, and there was no financial autonomy. This situation changed with the democratic Constitution of 1976, when an autonomous local self-government system was formally instituted and implemented.

According to the 1976 Constitution, Portugal is a unitary and decentralized state organized under the principles of subsidiarity, autonomy of local government, and democratic decentralization of the public service.[1] The Constitution instituted a new system of local self-government, a system with three tiers—administrative regions, which has not yet been implemented,[2] municipalities, and parishes[3]—all of them with directly elected bodies[4] and with politico-administrative and financial autonomy.[5] In the case of Azores and Madeira, instead of administrative regions, the Constitution considered, for the first time, a form of regional political autonomy, establishing an autonomous region in each of the two archipelagos.[6] Although with a regional scale, the administrative region is a form of administrative decentralization, in the same way as municipalities and parishes are for lower geographical tiers. It should, therefore, not be confused with the autonomous region, in Azores and Madeira, which is a form of political decentralization, an advanced form of devolution from the state to sub-national tiers. The municipality and the parish are

[1]Assembleia da República (2005). Constitution 1976, Article 6. See also: Canotilho and Moreira (1993), and Miranda (2007).

[2]See: Barreto (1998), Caupers (2009), Oliveira (1996a, b), Sá (1989), Santos (1985), and Silva (2000a, b, c, d).

[3]On the parishes ('freguesias'), see: Santos (1995).

[4]Citizens can also interfere in the local governance through local referendum (Law no. 4/2000, 24 August 2000; law no. 3/2010, 15 December; Law no. 1/2011, 30 November: 'referendo local').

[5]For a more detailed description and analysis of this process, see, among others: Silva (1995, 2004a, b).

[6]For the history of regional autonomy in Madeira, see: Carita (2007).

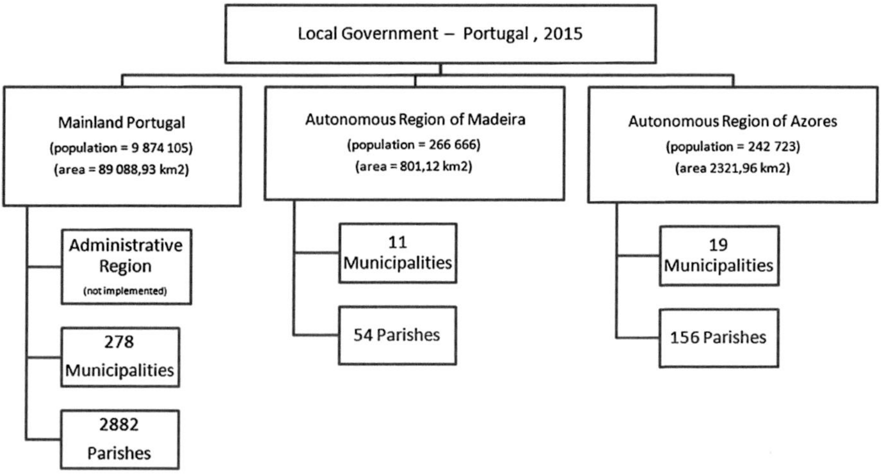

Fig. 2.1 The structure of local government in Portugal 2015

thus the only two tiers of sub-national self-government that cover the entire country[7] (Fig. 2.1).

The 1976 Constitution adopted the then existing 304 municipalities and 4025 parishes, inherited from the Estado Novo administrative organization, and defined broad principles, similar for all three layers of local self-government. Four new municipalities were created since then, and more than two hundred parishes were also created reaching the total number of 4260, a number that was reduced to 3092 in the 2013 parish merger reform implemented as part of the structural adjustment program (Tables 2.1 and 2.2).

The creation of new tiers of local self-government is not allowed, except in the case of the great urban areas and in the islands, where it is possible to introduce new forms of local self-government. In the case of the great urban areas, metropolitan institutions were created in 1991, in Lisbon and Porto, replaced in 2003 by the model of Great Urban Areas, Urban Communities, and Inter-municipal Communities, and applied this time to other parts of the country. This model was again altered in 2013.[8] All these institutional models were forms of inter-municipal cooperation or association, without directly elected boards.

[7]Since the end of decolonization process, in the mid 1970s, the territory of Portugal comprises the territory on the European mainland, the archipelagos of Azores and Madeira, the territorial waters, its exclusive economic zone, and the adjacent seabeds, as defined in Article 5 of the Constitution. According to the 1976 Constitution, Article 5 (7th Revision 2005), the state can only transfer any of the sovereign rights that it exercises over the national territory for the purpose of rectifying frontiers.

[8]Silva (2002a), and Silva and Syrett (2005, 2006) provide a critical perspective of the various attempts to implement a form of metropolitan government in Portugal. See also Moreira (2007) and Amorim (2009) on more recent developments on this issue.

Table 2.1 Local government in Portugal, 1974–2015

	Municipalities 1974	Municipalities 2015	Parishes 1974	Parishes 2015
Total	304	308	4025	3092
Mainland	274	278	3833	2882
Azores	19	19	139	156
Madeira	11	11	53	54

Source INE, census and statistical yearbook

Table 2.2 Municipalities and parishes: area and population

	Area (Km^2)	Population
1974*		
Municipalities	303.3	28,889
Parishes	22.9	2182
2015**		
Municipalities	299.4	33,712.6
Parishes	29.8	3358.2

Source INE, census and statistical yearbook
*Population data: 1970 census; these 1970 figures do not include the then colonial territories in Africa, Macau, and Timor; **population data: 2011 census

In the year that marks the 40th anniversary of the first democratic local election, held in December 1976, and when the outcome of the October 2015 legislative election marks the start of a new political cycle,[9] a critical assessment of the political and administrative decentralization process in Portugal is necessary. The chapter aims to contribute to this appraisal.[10] For that, the chapter addresses three main research questions: To what extent have the changes introduced in the local government system, during the last four decades, improved local democracy, and

[9]After the 2015 October election, the coalition PSD and PP formed a new government (XX Government, Decret-Law no. 249-A/2015, 9 November 2015; and Resolução do Conselho de Ministros no. 90-B/2015, 9 November 2015), which lasted only few weeks, as its program was rejected in Parliament (Moção de Rejeição no. 1-A/2015, 11 November 2015). A new government was then formed by the Socialist Party ('Partido Socialista') with the parliamentary support of the Communist Party (Partido Comunista Português), the Ecologist Party (Partido Ecologista 'Os Verdes'), and the Left Bloc (Bloco de Esquerda). The new prime minister and his government were nominated by the President of the Republic on November 26, 2015 (Decreto do Presidente da República no. 129-C/2015, 26 November 2015). The XXI Government structure was approved by Decret-Law no. 251-A/2015, 17 December 2015. The council of ministers is regulated by Resolução do Conselho de Ministros no. 95-A/2015, 17 December 2015.

[10]The chapter is part of an ongoing research interest focused on the local government system and policies in Portugal in the II Republic (1974–…), which is expressed, among other publications, in: Silva (1995, 1996, 2002a, b, 2004a, b, 2006, 2009, 2014, 2015a, b). Parts of the analysis presented in this chapter, in particular for the first decades of the II Republic, is taken from some of my previous publications listed here. The chapter adds new evidence and insights for the most recent period considered in the analysis.

increased local autonomy? How far did the changes introduced during the 2011 structural adjustment program represent a counterrevolution against the local government system introduced by the 1976 Constitution? Can the XXI Government' proposals for local government become a turnaround in the policies applied to local government during the bailout program? To answer these questions, the research on which the chapter is based uses a qualitative case study approach. It uses multiple types of data, mainly key local government policy documents and national legal acts pertaining to local and regional government organization and policies.

By answering these questions, the chapter offers a critical scrutiny of this complex process of political and institutional reform, which is marked by considerable positive changes, when compared to the non-democratic period that lasted between 1926 and 1974, although there are still important issues pending in central–local relations. Two examples of this are, for instance, the fact that local self-government continues to be responsible for a low proportion of public revenues and public expenditure, and the fact that a proper tier of regional self-government in mainland Portugal, the administrative region, has not yet been implemented, four decades after the approval of the 1976 Constitution.

The following sections explore and discuss the three research questions, addressing the issue of local autonomy, the implementation of a regional self-government tier, and forms of inter-municipal cooperation.

2.2 Local Autonomy: Continuity and Change

2.2.1 The 1976 Constitution: The Guarantee of Local Autonomy

The current local government system is based on the 1976 Constitution. However, municipalities are a very old form of local administration that goes back to the medieval period.[11] Parishes correspond also to a very old form of organization, although being only a division within the organization of the Catholic Church until the liberal period in the nineteenth century.[12] In both cases, the organizational structure comprises an elected assembly with decision-making powers and a collegial executive organ that is accountable to the assembly, which in the case of the municipality is also directly elected. Assemblies are elected by proportional representation, in universal, direct, and secret suffrage of the citizens who are registered to vote in the respective area. The municipality has two representative organs, the municipal assembly, with deliberative powers, and the municipal council, the

[11]For an historical overview of local administration in Portugal, see: Coelho and Magalhães (1986), Moreno (1986) and Oliveira (1996a, b).

[12]See: Santos (1995).

collegial executive organ, elected for a 4-year term.[13] Municipalities can create consultative boards, as is the case of the Youth Municipal Council, the Education Municipal Council, or the Municipal Security Council.[14] In the case of the parish, the assembly is the decision-making organ, and the parish council, the collegial executive organ. In parishes with a very small population, the parish assembly is replaced by the plenary meeting of registered electors. All municipalities have the same status, competencies, and administrative powers, no matter the demographic size of the municipality, and the same happens with the parishes.[15]

2.2.2 *Municipal Competences: Expansion, Diversification, and Centralizing Trends*

The competences of the municipalities expanded over the years, since 1976, although in some cases the transference or delegation of new functions, in different sectors (e.g., education, civil protection, health, social housing, justice, and road infrastructure), did not mean autonomy in the execution, since some of them corresponded to social obligations of the state. Besides the overall expansion of its functions, the profile of these functions and competences did also change. From an overwhelming dominance of infrastructures in the first decades, as a result of the country poor level of infrastructure in the beginning of the democratic period in 1974, the profile of municipal activities has been moving towards including increasingly more social functions, associated with the increasing role municipalities

[13]For a discussion of the recent changes in the local political leadership model, see Silva (2009). The political minority in the elected local government boards have the right to develop a democratic opposition, according to Law no. 24/98, 26 May 1998 ('Estatuto do Direito de Oposição'). Replaced Law no. 59/77, 5 August 1977. In 2005 was introduced legislation to limit the number of years in office for local councillors (Law no. 46/2005, 29 August 2005—'estabelece limites à renovação sucessiva de mandatos dos presidentes dos órgãos executivos das autarquias locais').

[14]Law no. 8/2009, 18 February 2009—Youth Municipal Council ('define o regime jurídico dos conselhos municipais de juventude'); Decret-law no. 7/2003, 15 January 2003—Education Municipal Council ('define os conselhos municipais de educação'); Law no. 33/98, 18 July 1998—Municipal Security Council ('define os conselhos municipais de segurança').

[15]The structure, competences, and the functioning of municipalities and parishes are regulated by the new local government act—Law no. 75/2013, 12 September 2013. This law represents to some extent a change in several aspects of the local government system. For a critique of this law and the process behind it, see Alexandrino (2014). It is expected to be changed again as part of the local government reform announced by the XXI Government (2015–2019). See Fonseca (2013) for a discussion of the proposed system of transference of competences for the parish (Proposta de Lei no. 104/XII).

have in the social area.[16] In 'Portugal 2020,' the fifth European support framework applied to Portugal, since the country became member of the European Community in 1986, there is a substantial shift in the paradigm of the European financial support. If in the previous frameworks, the emphasis was on the basic infrastructures, in the 'Portugal 2020,' resources will be applied mostly in the economy (competitiveness, internationalization, and employment) and in social inclusion, which means increasing responsibilities in new areas for the municipality, namely in the field of social inclusion. This change in the paradigm of EU financing will accelerate the ongoing changes in the profile of local government activities. In this context, it is important to implement new municipal planning instruments, such as the Social Chart,[17] which can be municipal or inter-municipal, a critical instrument for planning the increased competence municipalities now have in the social area (e.g., education, social housing, health, culture, and sports) and the Local Councils for Social Action[18] as well.

Paradoxically, the expansion and diversification of municipal functions seem to be followed by a centralizing trend in the way new functions and competences are assigned to the municipalities. As argued in Silva (2015a, b: 245), 'there is evidence that suggests a centralising trend in the context of new managerial practices, in which central government retains an important control role, by setting up the core values and priorities, for example, within the EU funding schemes, or in central-local partnerships at the municipal level, which is similar to what has been found in other European countries.' This pattern became even worse during the application of the MoU, as claimed by the national association of municipalities on several occasions.[19] In sum, there is a gap between the political discourse, clearly favorable to an increased decentralization and reinforcement of local autonomy, and the practical outcomes of the multiple reforms made in the local government system

[16]There are different perspectives in relation to how municipal competences should be defined. The principle of generality offers more guarantees for local government autonomy, although this seems not to be the perspective of the national association of municipalities or of part of its members as expressed in the XXII Congress (ANMP 2015). But I do agree with the ANMP that it is necessary an additional act defining the limits between what is responsibility of municipalities from what belongs to central government, as was done in 1984 and to some extent also in 1999. See Fonseca (2013) for a discussion of the proposed system of transference of competences for the parish (Proposta de Lei no. 104/XII). The new local government act ('Regime Jurídico das Autarquias Locais—Law 75/2013, 12 September) defines the competences of municipalities by the principle of generality, which is in accord with the Constitution, contrary to the previous Local Government Act (Law 159/99, 14 September 1999), which opted for a more detailed enunciation of these competences. The Decret-law 30/2015, 12 February, defines the delegation of competences from central government to the municipalities and inter-municipal entities. Nonetheless, this is clearly not a process of decentralization as a delegation is an ephemeral process, potentially unequal, and one that does not lead to the true reform of the state.

[17]Carta Social.

[18]Conselhos Locais de Ação Social.

[19]For instance, in the conclusions of the ANMP XX Congress, in 2012 (ANMP 2012a, point 6).

in this period.[20] The share of local government in the public administration sector, given by the percentage share of local government in the total public revenue or expenditure, which continues below European levels, is a good indicator of the level of administrative centralization.

This centralizing trend can also be observed in other dimensions of the local government system: in the organization of its own services or in the local finance system. For instance, the organization of the municipal services changed in 2009, after 25 years.[21] During the application of the MoU in 2011 and following years, municipalities saw their autonomy in the organization of their own services and staff deeply affected, becoming dependent of central government decisions.[22] This was partially altered in 2014, as part of the agreement the ANMP and central government reached regarding the 'Fundo de Apoio Municipal.'[23] A similar pattern emerges in the local finance system. Altered several times since the first local finance law was adopted in 1979, without putting in question the essence of the local finance system and the respective municipal finance autonomy, the local finance system was affected in the recent years by decisions that tend to put in danger local autonomy.[24] For instance, as the ANMP argued on May 2011, when the XVIII Government was overthrown, municipalities had already suffered cuts of around 905 million Euros since 2005, being for that reason in the limit of their

[20]Important to note here, although outside the aim of this chapter, is the gradual move, in some areas or sectors of municipal policy, from traditional modes of hierarchical government to modes of networked governance (see, e.g., Silva 2004a, b).

[21]Decret-Law no. 116/84, 6 April 1984 replaced by Decret-Law no. 305/2009, 23 October 2009.

[22]Law 49/2012, 29 August—new rules for municipal staff (senior officers) similar to those applied in central government (Law 64/2011, 22 December).

[23]ANMP (2014c). Comunicado 'Acordo ANMP—Governo sobre o Fundo de Apoio Municipal,' Coimbra, July 8, 2014. For an analysis of the FAM (Fundo de Apoio Municipal) see Silva and Santos (2014). This municipal support fund (Law 53/2014, 25 August) is the instrument designed to support the new regime (PAM—Municipal Adjustment Program) intended to provide financial assistance to the municipalities in a situation of financial rupture, according to the local finance act (Law 73/2013, 3 September). The municipalities engaged in this program are subjected to several limitations and to mandatory actions, which clearly undermines the autonomy of local government.

[24]Law 1/79; Decret-law 98/84; Law 1/87; Law 42/98; Law 2/2007; Law 73/2013. See Rocha and Pinto (2014) for an analysis of the different local finance acts. In practice, there was a decline in the share of local government in the national tax income after the first revision of the Law 1/79, an effect particularly important in small and medium-sized municipalities, namely those in the interior of the country and in the Azores and Madeira. The approval of Law 2/2007 motivated an extraordinary congress of the ANMP in 2006, in which the new law was rejected by the national association of municipalities.

financial capacity.[25] Legislation published during the MoU period, as the law on municipal debt[26] or the 2013 local finance act,[27] increased the financial difficulties, making the municipal financial situation worse than that experienced under the Stability and Growth Programme (PEC)[28] implemented in the years before the bailout program by the XVII and XVIII Government. In 2012, the volume of financial transfers from the state to the municipalities was equal to that of 2005, which represented a severe pressure on local government budgets (ANMP 2012a).[29]

The Local Economy Support Program (PAEL) placed even more pressure on the municipalities engaged in this program.[30] With the argument that it was necessary to regularize all short-term debts, the program forced the municipalities involved in that assistance program to increase to the maximum all municipal fees and taxes,[31] compromising during several years the financial autonomy of all municipalities that adhered to the program. In 2014, the ANMP reached an agreement with central government, concerning the conditions of the FAM.[32] The terms of this agreement were, for the ANMP, a better solution for the municipalities in need of financial

[25]ANMP statement, May 4, 2011 (Comunicado 'Memorando de Entendimento da Troika: ANMP Contra a Proposta Para Redução dos Municípios,' Lisbon, 4 May 2011). The ANMP refers that the municipalities had together a positive result of around 70 million Euros in 2010. In 2011, the 'superavit' was nearly 196, 1 million Euros. The total municipal debt had also a positive evolution in these years and represented only 4 % of the total public debt; the state debt is 96 % (ANMP 2012a).

[26]Law 8/2012, 17 March [Lei dos Compromissos e dos Pagamentos em Atraso (LCPA)]. And Decret-law 127/2012, 21 June. It was required by the 'Memorandum of Understanding on specific economic policy conditionality (MoU).' Revised in 2015 by Law 22/2015, 17 March and Decret-law 99/2015, 2 July.

[27]Law no. 73/2013, 3 September 2013 (defines the finance system of municipalities, parishes, and inter-municipal entities).

[28]PEC—Programa de Estabilidade e Crescimento (Stability and Growth Programme): Programa de Estabilidade e Crescimento 2005–2009; Programa de Estabilidade e Crescimento 2006–2010; Programa de Estabilidade e Crescimento 2007–2011; Programa de Estabilidade e Crescimento 2008–2011; Programa de Estabilidade e Crescimento 2011–2014.

[29]The MoU established the need to reduce at least 175 million Euros in the annual budget transfers from the state to the local and regional administration. This requirement leads to the revision of the Local Finance Act (municipalities and parishes), as well the Regional Finance Act (for the 2 autonomous regions).

[30]The PAEL (Programa de Apoio à Economia Local) was approved in August/September 2012 (Law 43/2012, 28 August; Portaria 281-A/2012, 14 September). The program consisted in medium-/long-term loans to be used to pay municipal debts with more than 90 days on March 31, 2012. The program approved the proposals of 112 municipalities. Only in 2013 did the municipalities included in the program receive this support. Only 6 municipalities received a first parcel in the last week of 2012.

[31]ANMP (2013a). Comunicado 'A ANMP e as Taxas de IMI,' Coimbra, May 17, 2013.

[32]Law 53/2014, 25 August: FAM—Fundo de Apoio Municipal ('Municipal Support Fund').

assistance than the version initially approved by central government in June 2014.[33]

Local government can create municipal and inter-municipal enterprises to fulfill its competences in different sectors, an area that has also been affected during the implementation of the structural adjustment program.[34] Municipalities, namely through its national association, expressed, on numerous occasions, opposition to privatization of public services in sectors considered sensitive. It was the case of the reorganization and privatization in the urban waste sector.[35]

Numerous other decisions taken by central government contributed to this concentration trend in central–local relationship in Portugal. It is the case, for instance, of the following decisions: inclusion of policy measures in the Memorandum of Understanding (MoU) between Portugal and the international institutions with impact on the local autonomy without negotiation with the municipalities and parishes[36]; definition of the new EU funding framework ('Portugal 2020') without appropriate negotiation with the municipalities[37]; adoption of legislation that clearly limits the municipal administrative and financial autonomy, as was the case of the law on the payment of municipal debts, the law on the number of senior municipal staff, the changes in the organization of municipal enterprises, setting up ratios that forced the closure of some of them, despite the fact that in some cases, these organizations fulfilled important social objectives and

---

[33]ANMP (2014a, b, c, d, e, f). Comunicado 'Acordo ANMP—Governo sobre o Fundo de Apoio Municipal,' Coimbra, July 8, 2014. As part of this agreement, the government agreed the following: (1) to revise the 'Lei dos Compromissos e Pagamentos em Atraso'; (2) to devolve the autonomy municipalities had previously in the organization of their services and in the management of their staff, having as the sole limit the total amount of salaries paid on average in the last 3 years; (3) to guarantee that the process of reorganization of the water and sewage sector would not involve the privatization of the water sector; and (4) to include incentives for municipalities in the green taxation reform being prepared at that time. The ANMP considered this was not the ideal agreement but was one that respected and reinforced the municipal autonomy. For an analysis of this new version, see Gomes (2015).

[34]Law no. 50/2012, 31 August 2012 defines the regime of these local public enterprises. Replaced Law no. 53-f/2006, 29 December and Law no. 55/2011, 15 November. In the case of the water, waste, and sewage, municipalities can create special services (Decret-law no. 194/2009, 20 August 2009—'*estabelece o regime jurídico dos serviços municipais de abastecimento público de água, de saneamento de águas residuais e de gestão de resíduos urbanos*'). In the same area/sectors can be created multimunicipal systems, defined in Decret-law no. 92/2013, 11 July 2013—'*define o regime de exploração e gestão dos sistemas multimunicipais de captação.'tratamento e distribuição de água para consumo público, de recolha, tratamento e rejeição de efluentes e de recolha e tratamento de resíduos sólidos*'. Silva (2000c) examines the introduction of these entities in the local government system in the late 1990s and early 2000s.

[35]It was the case of Lei no. 45/2014 de 20 de Março. Municipalities and its national association opposed the privatization of the public enterprise 'Empresa Geral de Fomento.' See, among other, ANMP (2014b, d).

[36]As referred in the conclusions of the XX ANMP Congress (2012a, point 8).

[37]ANMP (2014a, b, c, d, e, f). Comunicado—'ANMP considera "inaceitável e intolerável" não ser chamada a participar na elaboração dos regulamentos do Portugal 2020.' The same happened with other important issues, as was the case of changes associated with the 'green taxation' (ANMP 2014a).

when similar criteria were not applied to public enterprises controlled by central government.

This concentration trend is also present in the following decisions: the approval of laws that gave the regulatory entity for the water, sewage, and urban waste sector competences and powers that typically belong to local government[38]; the annual definition of limits to the municipal expenditures with staff; the publication of norms forbidding the admission of new staff for the municipalities, as well as norms forcing the reduction in the number of municipal employees; the lack of transparency in some contracts celebrated between the state and some municipalities; the reduction in municipal revenues, with no respect for local autonomy given by the Constitution; the annual definition in the state budget of debt limits for the municipalities, different from those established in the local finance law, as well as penalties for those exceeding these limits; the exemption of payment of municipal taxes and fees by the state, while the reverse is not applied; the reform of the judicial map without consultation with the municipalities[39]; the attempt to create new forms of local government, later considered unconstitutional, being rejected by the Constitutional Court.

Also in the field of spatial planning, this concentration trend was present: the substitution of municipalities by the CCDR in certain land-use planning licensing decisions; the management of harbor areas independently of the municipal competences in land-use planning; the criteria for the de-classification of national roads with which the municipalities did not fully agree during the negotiations[40]; exemption of municipal licenses for building works carried out by the state.

In addition to the concentration trend identified in the decisions enumerated before, it is also unacceptable the difference that continues to exist between the municipalities and parishes in mainland Portugal and those in the two autonomous regions. Numerous competences that in mainland Portugal have been transferred

[38]Entidade Reguladora dos Serviços de Águas e Resíduos (ERSAR). See: ANMP (2014b). Comunicado sobre as propostas de Lei que estabelecem o regime jurídico dos serviços municipais de abastecimento público de água, de saneamento de águas residuais urbanas e de gestão de resíduos urbanos. Coimbra, 5 February 2014.

[39]ANMP (2012a, b). Comunicado 'Municípios têm de ser ouvidos na questão do encerramento de Tribunais Judiciais,' Coimbra, January 31, 2012; ANMP (2013b). Comunicado 'A ANMP e o mapa judiciário ou a deslocalização e encerramento de mais serviços públicos de interesse geral,' Coimbra, October 23, 2013. The ANMP opposition to this policy included others forms of public expression (e.g., the presence of over 100 councillors in the Parliament during the plenary discussion of this legislation, on May 2, 2014, as a form of protest; regional meetings of municipal councillors to discuss this and other issues related to the governmental decisions derived from the implementation of the MoU, such as reorganization and closure of other local public services, other than tribunals, namely primary schools, the PAEL—Programa de Apoio à Economia Local, and the Fundo de Apoio Municipal). See also the conclusion of the XXI ANMP Congress (ANMP 2013d).

[40]ANMP (2014e). Nota do Conselho Diretivo Proposta de Diploma: 'Estatuto das Estradas da Rede Rodoviária Nacional,' Coimbra, September 2014.

and assigned to the municipalities and parishes, in the autonomous regions, are in the hands of the respective regional government, thus configuring an unequal and discriminatory situation.

2.2.3 *The 2013 Parish Merger Reform: Disrespect of Local Autonomy*

Municipalities and parishes can be established and abolished, and their areas can also be changed, by law, after consultation with the local authorities affected. Contrary to what happened in several other European countries in the 1960s and 1970s, there was no similar large merger reform of municipalities or parishes in Portugal in the second half of the twentieth century. Nonetheless, a similar large-scale merger reform took place much earlier, in the first half of the nineteenth century, a process that was followed by a series of readjustments, which ended in an even smaller number of municipalities at the beginning of the twentieth century. There are currently 308 municipalities, four of which created after 1974. In 2013, there were 4260 parishes, 234 of which created after 1974, a number that was substantially reduced as a result of the 2013 parish merger reform, from 4260 to 3092,[41] a reform forced by the Memorandum of Understanding on specific economic policy conditionality (MoU), signed between the Portuguese Government and the EU Commission, the European Central Bank, and the International Monetary Fund, in 2011.[42]

Although the MoU did not define the procedure to be followed, the government rejected the possibility of a local referendum, as the European Charter of Local Autonomy recommends every time a territory is changed, and did not give the due importance to the opposition expressed by numerous local government units and the respective national association. Nonetheless, the MoU was quite clear about the need to reorganize local government administration: «central government should develop

[41]Based on Law no. 22/2012, 30 May 2012; Law no. 11-A/2013, 28 January 2013 (administrative reorganization of the parishes' territory); and Law 81/2013, 6 December. The national association of municipalities (ANMP) was against this process, as clearly stated in its XX Congress in 2012. On the contrary, it was in favor of the reinforcement of the metropolitan areas and inter-municipal Communities and creation of administrative regions. For the ANMP, the fusion or aggregation of parishes should be done based exclusively on the initiative of each local authority or the respective local populations. Parallel to the national reform in the number of parishes, the municipality of Lisbon was already conducting a reform of its parish administrative division, which was approved by Law no. 56/2012, 8 November 2012 (administrative reorganization of Lisbon), reducing the number of parishes from 53 to 24 (Belém, Ajuda, Alcântara, Benfica, São Domingos de Benfica, Alvalade, Marvila, Areeiro, Santo António, Santa Maria Maior, Estrela, Campo de Ourique, Misericórdia, Arroios, Beato, São Vicente, Avenidas Novas, Penha de França, Lumiar, Carnide, Santa Clara, Olivais, Campolide, Parque das Nações).

[42]Portugal: Memorandum of Understanding on specific economic policy conditionality. 3 May 2011 (EC/ECB/IMF 2011; Silva 2014).

until July 2012 a consolidation plan for the reorganization and significant reduction of the number of municipalities and parishes, in articulation with EC and IMF staff. This reform was expected to enhance service delivery, improve efficiency, and reduce costs». Only in the case of parishes was the Portuguese government able to reduce the number of local authorities. There were no changes in the number of municipalities, and it is implausible that it will happen in the near future. The National Association of Municipalities expressed on several occasions complete opposition to the proposed reduction in the number of municipalities.[43, 44, 45]

2.2.4 *The 2015 Political Shift: The Expected Impact on Local Autonomy*

The political shift that followed the October 2015 legislative election is expected to impact on the organization and competences of municipalities and parishes. After 4 years of austerity, also imposed on municipalities and parishes, and after the implementation of the parish merger reform, the XXI Government proposals point for a reversal of some of the changes introduced in the context of the Memorandum of Understanding on specific economic policy conditionality (MoU), implemented since 2011 by the XIX Government (2011–2015).

Shortly after taking office, the XXI Government announced the beginning of the proposed local government reform, clarifying the model and the calendar of its decentralization plan.[46] The aim is to expand local democracy, to improve local public services, and to give new competences to local government (Governo de Portugal 2015). The plan requires the revision of the CCDR competences as well as those of the metropolitan areas, reinforcing the respective democratic legitimacy. The process engages 10 ministries and is expected to be concluded in mid 2017, in

[43]ANMP statement, May 4, 2011 (Comunicado 'Memorando de Entendimento da Troika—ANMP Contra a Proposta Para Redução dos Municípios,' Lisbon, May 4, 2011). Three main reasons were referred: First, the extinction of municipalities is a competence of the Parliament not a governmental competence; second, such reform should be considered within a larger process involving also the reform of the state; and third, the Portuguese municipalities are among the largest on average in Europe, whatever the criteria used, including the demographic one.

[44]For a perspective from inside the political process behind this reform of local administration, see Relvas and Júlio (2015). The authors of this book are the former minister and secretary of state responsible for the reform of local administration during the XIX Government. For a discussion about the need of this reform, see: Oliveira (2011). See also the document (Green Book) prepared by the XIX Government and on which the reform of local administration was based: Governo de Portugal (2011). And Carneiro (2012) for a discussion of these proposals.

[45]For an analysis of the impacts of the MoU on other dimensions of local government, see Alexandrino (2012).

[46]Comunicado do Conselho de Ministros, 14 January 2016.

order to allow the new local government electoral cycle (2017–2021) to develop within this new political and institutional framework.

The municipal competences in specific areas will be reinforced, namely in the education, health, social services, transport, culture, housing, civil protection, public security, ports, and maritime areas. Parishes will have their own competences in areas in which currently they only have powers delegated by the municipality. They will have competences differentiated according to their nature. Another important consequence of this proposal for the reinforcement of local government powers and competences was the cancelation of the privatization of public transport in Lisbon and Porto, one of the first legislative initiatives taken by the new political majority in Parliament.[47] The evaluation of the 2013 parish merger reform implemented by the XIX Government is one of the other key proposals for the reform of local government announced by the XXI Government. This will be done in close dialogue and cooperation with all parts concerned. In connection with this are the proposed development of inter-municipal cooperation and the democratization of metropolitan areas through the direct election of its boards. The reform of the local government finance system, with the aim to reach the European average of participation in public revenues, will represent, if fully implemented, a major improvement in the financial capacity of local government. This will be done by increasing the participation of municipalities in the main national taxes,[48] among other measures. Besides the initiatives to be taken by the XXI Government, according to its program, other parties have already taken initiatives that may well have positive impacts on the autonomy and capacity to act of municipalities and parishes. One example is the first project of law presented in the new parliament,[49] which is intended to increase the number of full-time members of the Parish executive board. The fact that only the president can be full time has prevented in the past the decentralization of more competences to this lower tier of local self-government.

[47]For instance, several project laws presented, in November 2015, by political parties that support the XXI Government: PCP (project-law 23/XIII; 25/XIII) and BE (project-law 47/XIII; 48/XIII; 49/XIII; 50/XIII), concerning the 'privatization' of public transport in Lisbon and Porto—Carris, Lisbon Metro, STCP, and Porto Metro.

[48]IRS (Personal Income Tax), IRC (Corporation Income Tax), and IVA (Added Value Tax).

[49]Projeto de lei no. 119/XIII (1.ª), 30 January 2016: Procede à alteração do regime de permanência dos membros das Juntas de Freguesia (presented by the party PAN—Pessoas, Animais, Natureza). This proposal will change the current system defined in: Article 27.º, Law no. 169/99, 18 September 1999; and in the Article 5.º and Article 7.º Law no. 11/96, de 18 April 1996. The statute of local councillors is regulated by Law no. 29/87, 30 June 1987 ('Estatuto dos Eleitos Locais'), changed by Law no. 97/89, 15 December, Law 1/91, 10 January 1991; Law 11/91, 17 May 1991; Law 11/96, 18 April 1996, Law 127/97, 11 December, Law 50/99, 24 June, Law 86/2001, 10 August 2001; and Law 22/2004, 17 June 2004, changed and republished by Law no. 52—A/2005, 10 October.

2.3 The Regional Tier

2.3.1 The 1976 Constitution: An Unfinished Institutional Revolution

The regional layer of public administration between the state and municipalities has a long history in Portugal.[50] The current system is defined in the 1976 Constitution, comprising forms of decentralized institutions—autonomous regions and administrative regions—as well as de-concentrated institutions.

In the case of administrative de-concentration, it has been common practice during the last four decades to have central government ministries organized in regional and/or local de-concentrated tiers,[51] although based on different regional divisions according to the specificities of each sector. Among them is particularly relevant the case of the five regional planning and coordination boards, the CCDR, one for each of the five NUT-II, namely because they have been associated with the creation of the future administrative regions, and expected to be the technical structure that will support the future administrative regions.[52] Some ministries changed their regional organization, in some cases more than in one occasion since 1974, although in the recent years, there has been a tendency for central government de-concentrated departments to follow the five planning regions' map, coincident with the five NUT-II units.

Political and administrative decentralization includes the two autonomous regions of Azores and Madeira, a form of political decentralization, whose boards were first elected in 1976, and the administrative region, a form of administrative decentralization, due to be instituted only in mainland Portugal. The option for a form of political decentralization in the archipelagos of Azores and Madeira, instead of a mere administrative decentralization, as in the case of the administrative region, was somehow the outcome of the recognition of the specific geographical, economic, social, and cultural characteristics of these two archipelagos, as well as of

[50]For a detailed perspective of this historical process, namely in the first decades after the adoption of the 1976 Constitution, see the texts quoted in note 2. See also Carita (2007) for the case of Madeira.

[51]For instance, the case of the ministries of health, education, agriculture, environment, and so on.

[52]CCDR—Comissão de Coordenação e Desenvolvimento Regional. The boundaries of these regional divisions have changed over the years. They were first created in the last years of the Estado Novo, as part of the 5-year development program, being renamed several times since then, besides the adjustments in their geographical areas and in its competencies. CCDR is now an institution more open to the participation of different local interests. For an historical overview of this process, see: Oliveira (2008).

the historic aspirations for autonomy of the population of that part of the national territory, as is clearly stated in the 1976 Constitution.[53]

Administrative regions have two representative organs[54]: the regional assembly, the region's decision-making organ, and the regional council or regional government, the collegial executive organ. The regional assembly is the region's deliberative organ and is elected by universal, direct, and secret suffrage according to the principle of proportional representation. Each region has a representative of central government. The creation of administrative regions was postponed on several occasions, and a proposal for the creation of eight regions, put forward by the XIII Government, supported by the Socialist Party, was rejected in the 1998 national referendum.[55] Nonetheless, despite the negative result in the referendum, the administrative region was not removed from the Constitution. But it had in the following years a much lower priority in the political discourse than before.

In the two right-centered coalition governments that followed the emphasis was increasingly put on the reorganization and reinforcement of the de-concentrated regional administration. Even in the XVII Government (2005–2009), the socialist government that followed, the creation of administrative regions was dependent of a national referendum and should take into consideration the five planning regions in mainland Portugal, corresponding to the NUT-II (PCM 2005). This should be complemented by a reorganization of the de-concentrated departments of central government, according to the same regional map, as well as the reinforcement and stabilization of the CCDR functions, namely the coordination of central government policies at the regional scale. This reinforced the idea that the creation of administrative regions should be based on the five planning regions and should be implemented after the reform of the de-concentrated central government structure

[53]See Carita (2007) for the case of Madeira. Despite the differences among political parties, on some key points it is possible to talk of a national political consensus in favor of the regional autonomy in Azores and Madeira. This political commitment in favor of the regional autonomy is well expressed in the 2015 electoral manifesto of the Socialist Party ('Partido Socialista, 2015, pp. 48.'), the party that supports the current XXI Government (2015–2019). The same can be found in the 2015 electoral program of the coalition between the Social Democratic Party and the Popular Party (Coligação Portugal à Frente 2015, pp. 129)—'Agora Portugal Pode Mais—Programa Eleitoral—the main opposition political block in the current parliament (2015–2019). This political commitment in favor of the regional autonomy is also expressed in the 2015 electoral manifesto of Bloco de Esquerda (2015, pp. 59–64)—'Manifesto eleitoral–Bloco de Esquerda, Legislativas 2015'—and in the 2015 electoral program of the Communist Party ('Partido Comunista Português, 2015', pp.70). Finally, it is also expressed in the program of the XXI Constitutional Government (2015–2019), pp. 85–86.

[54]Law 56/91, 13 August established, among other aspects, the organization and competences of the regions. It was approved by unanimity in Parliament. The Law 19/98, 28 April, on the creation of regions was approved by an overall majority.

[55]The literature on the attempts to institutionalize the administrative regions in Portugal after 1974 is extensive. For an historical overview of the first decades of this process, see, among others, the texts quoted in the note 2; see also, MEPAT (1998). Syrett and Silva (2001) examine the creation of regional development agencies, one of the various attempts made over the years to create regional-wide institutions.

according to these five regional divisions. The essence of this political option was maintained by the XVIII Government (2009–2011), supported by the Socialist Party, although without any practical consequence, due to the economic and public finance crisis that caused the downfall of this Government, in the middle of its mandate, which was due to last until 2013.

The election of a new political majority supported in parliament by the Social Democratic Party and by the Popular Party, and the implementation of the structural adjustment program (MoU) by the XIX Government (2011–2015), changed considerably the terms of this debate. The most important single fact in this regard was perhaps the withdraw of regionalization from the list of political priorities. Nonetheless, the national association of municipalities continued to claim the implementation of administrative regions and proposed, once again, in its XXII national congress, held in 2015, the implementation of the administrative regions.

2.3.2 *The 2015 Political Shift: Half Way Between the Constitution and the Status Quo?*

The outcome of the 2015 October legislative election may lead to a turning point on the current status quo although not a move exactly to what is in the 1976 Constitution. The Socialist Party's 2015 electoral manifesto and the XXI Government's program point clearly to the reform of the current regional administrative structures as well as those in the metropolitan areas. The new territorial model proposed will create five planning and development regions with elected boards. These new entities will be based on the current CCDR structures and will adopt the same territorial division. The executive board will be elected by an electoral college formed by the members of the municipal executive and deliberative boards from all municipalities in the respective region. This executive with 3–5 members will respond to the regional council. Contrary to the previous XVII and XVIII Governments, supported by the Socialist Party, the proposed creation of regions, this time with indirectly elected boards, will not require a national referendum. If it is not exactly what had been considered in the past, and if not exactly what is in the Constitution, it is clearly an improvement in the local democracy compared to the current situation, as it democratizes the existing regional administrative structures. It is to some extent a compromise between a simple re-organization of the de-concentrated regional administration and a fully new regional organization with directly elected boards. Part of the reform of the CCDR, besides the democratization through the election of the president of the executive board, is the reform of the de-concentrated state administration and the subsequent integration of some of these services in the new regional planning and development entities that will replace the current CCDR.

2.4 Inter-municipal Cooperation: The Way for Decentralisation?

Municipalities can also develop other models of inter-municipal cooperation, namely municipal associations, with specific responsibilities and competences, in order to administer common interests, which they have done consistently and with positive results in the last decades.[56] Due to the nonexistence of administrative regions, the multiple forms of inter-municipal cooperation implemented in the last decades have been a key player in central–local relations in mainland Portugal. On numerous issues, they act, occasionally, as a de facto regional or sub-regional representative institution.[57] Municipal associations and other forms of inter-municipal cooperation became even more important in the last years, in particular when, in the aftermath of the negative result in the 1998 national referendum, the creation of administrative regions was postponed for the following electoral cycles.[58] Besides the efforts made for the harmonization of central government regional divisions, as referred before, based on the NUT-II map, there have been efforts over the years to harmonize the boundaries of municipal associations, based on the NUT-III map.[59]

The model of metropolitan government introduced by Law 44/91 and altered in 2003 by Law 10/2003 and Law 11/2003, and in 2008 by Law 45/2008 and Law 46/2008 proved to be inadequate, due to the lack of direct political legitimacy, to the inexistence of own metropolitan competences, the difficulty to articulate metropolitan interests with the interests of each municipality in particular, and finally also because this form of metropolitan government lacked popular recognition and support. The institutional model adopted in the case of the metropolitan authority of transport has also been criticized by the municipalities.[60] The 2013 local government act defined different types of inter-municipal entities: two

[56]Law 45/2008, 27 August on the Inter-municipal Communities (CIM). The last reform of inter-municipal entities, carried out by the XIX Government, is defined in the Law no. 75/2013, 12 September 2013. For an analysis of this new regime of inter-municipal entities, see Gonçalves (2014). In 2012, a pilot study on the new competences and financing of these entities was published by central government (DGAL 2012).

[57]For the specific case of inter-municipal cooperation in the metropolitan areas ('metropolitan government'), see: Silva (2002a, b), Amorim (2009), and Moreira (2007).

[58]On municipal associations, see: Silva (1993, 2006), Lopes (2009), Moreira (2007), and Amorim (2009).

[59]NUT-II and NUT-III regions follow the European nomenclature.

[60]The two metropolitan transport authorities (Lisbon and Porto) were instituted by Law 1/2009, 5 January and were abolished by Law 52/2015, 9 June. Their competences were inherited by the two metropolitan entities (Lisbon and Porto).

metropolitan areas—Lisbon and Porto—and 21 Inter-municipal Communities.[61] The same government attempted to transform the statute of these entities, Inter-municipal Communities and metropolitan areas, into new forms of local self-government, which was rejected by the Constitutional Court.[62] Currently, there are 3 types of inter-municipal associations, according to the 2013 legislation: the inter-municipal community, the metropolitan area, and the municipal and parish association with specific aims or object.

The national association of municipalities proposed on several occasions the introduction of a new model of metropolitan government, with metropolitan wide competences (e.g., metropolitan strategic spatial planning, metropolitan mobility plan, environment, education, health, among others), with directly elected boards, as well as an adequate technical structure. The new territorial model proposed by the XXI Government includes this perspective. In fact, besides the changes in the five planning and development regions, it will also implement forms of self-government with the directly elected boards in the metropolitan areas of Lisbon and Porto. The metropolitan assembly will be elected directly, being the president of the executive board the number one in the winning list in the election for the metropolitan assembly. The other members of the executive board will be elected by the metropolitan assembly, based on proposal made by the elected president of the executive board. The new metropolitan areas, with directly elected boards, will have own competences, namely in the following sectors: transport, water, waste, energy, economy, and tourism promotion, as well as in programs of regional development. They will be responsible for the definition and implementation of the metropolitan ecological structures. In addition to these changes, the new territorial model will reinforce inter-municipal cooperation through the existing Inter-municipal Communities. These entities will act as an instrument of inter-municipal cooperation in articulation with the new regional entities, which will emerge from the democratization of the current CCDR, as well as with the new elected metropolitan entities.

2.5 Conclusion

If the overall balance between what is in the Constitution and the current local government system supports more the idea of compliance than the idea of disagreement with the fundamental constitutional principles, it is also manifest that this

[61]Inter-municipal Communities: Alto Minho, Cávado, Ave, Alto Tâmega, Tâmega e Sousa, Douro, Terras de Trás-os-Montes, Região de Aveiro, Região de Coimbra, Região de Leiria, Viseu Dão Lafões, Beiras e Serra da Estrela, Beira Baixa, Oeste, Médio Tejo, Alentejo Litoral, Alto Alentejo, Alentejo Central, Baixo Alentejo, Lezíria do Tejo, Algarve.

[62]ANMP (2013c). Comunicado 'Iniciativa da ANMP de suscitar a inconstitucionalidade do diploma sobre o estatuto das comunidades intermunicipais coroada de êxito,' Coimbra, May 29, 2013.

is more clearly so in some institutional and policy dimensions than in others and more so also in some periods than in others, during the four decades since the overthrow of the dictatorship in 1974 and approval of the democratic constitution in 1976. A clear example of noncompliance with the Constitution is the inexistence of a democratically elected regional self-government tier.

During the four decades of democracy since 1974–1976, there have been changes in the local electoral system, in the internal structure of municipalities and parishes, in the local finance system, to mention just some of the key dimensions of the local government system, but none of these changes or reforms represented a complete rupture with what was defined in the 1976 Constitution and in the legislation adopted in the years that followed its approval. A similar continuity trend characterizes the policy areas under the direct responsibility of municipalities in Portugal. The current system of local self-government continues based on the 1976 Constitution, as none of the constitutional revisions affected the level of local autonomy assigned initially to local government in the Constitution (Silva 2004a, b, 2006, 2009).[63] If the austerity policy adopted by the XIX Government (2011–2015) during the application of the Memorandum of Understanding on specific economic policy conditionality (MoU) undermined key dimensions of the local autonomy, the proposals of the XXI Government (2015–2019) suggest, if fully implemented, that the reduction in local autonomy experienced by municipalities and parishes will be somehow revert. In this sense, continuity with fluctuations more than rupture is what emerges as the key feature in an overall balance of the local government system in Portugal in these four decades, despite the seven constitutional revisions and the recent structural adjustment program.[64]

Despite the rupture with the previous authoritarian political regime, expressed in the 1976 Constitution, there are still important issues that require reform and change, namely in the field of central–local relations.[65] It is the case, for instance, of the low proportion of public financial resources allocated to local government when compared to the European average. But it is also the case that nearly four decades after the approval of the 1976 Constitution, the administrative region has not yet been implemented. This delay in the implementation of administrative regions is against what is written in the Constitution and is also against the widely accepted vision of the members of the Council of Europe.[66]

[63]The regime of administrative tutelage is currently regulated by Law no. 27/96, 1 August 1996 ('Regime jurídico da tutela administrativa').

[64]In 1982, 1989, 1992, 1997, 2001, 2004, and 2005 (CRP 2005). A new constitutional revision was expected to take place during the period 2009–2013, as announced by the two main political parties in Parliament. Nonetheless, the political crisis that lead to the overthrown of the XVIII Government in the middle of its mandate, in 2009, and the following economic adjustment program implemented by the XIX Government changed the terms of this issue.

[65]For an analysis and discussion of some of these issues, see: Silva (2009, 2015a, b).

[66]For more on this, see: Silva (2014, 2015a, b), Caupers (2009), Oliveira (1996a, b, 2008), and Sá (1989).

The evidence suggests that the political consensus, regarding local government and decentralization, that existed before the Memorandum of Understanding on specific economic policy conditionality (MoU), was somehow affected by key political decisions taken and implemented by the XIX Government (2011–2015), as well as, before that, by the XVIII Government in the context of the several Stability and Growth Programme (PEC)[67] adopted. In fact, since 2010, the economic and financial crisis, and the social austerity policy implemented by the PEC's and by the MoU signed in 2011, justified an increased centralism, in particular in the sectors with impact on the public deficit. This move toward centralization had clear negative effects in the local autonomy and in the share of local government in the total public revenues, as had the continuous changes in the legal frameworks. In reaction to this counterrevolution, which affected local government autonomy and local democracy, the proposals put forward by the XXI Government (2015–2019), and the electoral programs and electoral manifestos of the three parties that support this government in Parliament, point for a reversal of some of those institutional changes, increasing the levels of administrative decentralization, which will have positive impacts on urban governance, and improving the quality of the local democracy.

References

Alexandrino JM (2012) A administração local autónoma: situação atual e propostas de reforma apresentadas na sequência do Memorando da Troika. Direito Regional e Local 18:5–13

Alexandrino JM (2014) A lei de reforma da administração local e os seus grandes problemas. Questões Atuais de Direito Local 1:7–20

Amorim JP (2009) Breves reflexões sobre os novos regimes das Associações de Municípios e das Áreas Metropolitanas de Lisboa e Porto. Revista de Direito Público e Regulação 2:73–102

ANMP (2011) Comunicado Memorando de Entendimento da Troika: ANMP Contra a Proposta para Redução dos Municípios. Lisbon. 4 May 2011

ANMP (2012a) Comunicado Municípios têm de ser ouvidos na questão do encerramento de Tribunais Judiciais. Coimbra. 31 Jan 2012

ANMP (2012b) XX Congresso da ANMP (Extraordinário): Autonomia e Financiamento do Poder Local e Reorganização Territorial do Estado. Conclusões - Mais Poder Local, Melhor Portugal. Santarém, 29 de setembro de 2012

ANMP (2013a) Comunicado A ANMP e as Taxas de IMI. Coimbra. 17 May 2013

ANMP (2013b) Comunicado A ANMP e o mapa judiciário ou a deslocalização e encerramento de mais serviços públicos de interesse geral. Coimbra. 23 Oct 2013

ANMP (2013c) Comunicado Iniciativa da ANMP de suscitar a inconstitucionalidade do diploma sobre o estatuto das comunidades intermunicipais coroada de êxito. Coimbra. 29 May 2013

ANMP (2013d) XXI Congresso da ANMP - Mais Governo Local, Melhor Portugal - Linhas Gerais de Atuação. Santarém, 23 de novembro de 2013

ANMP (2014a) Comunicado: ANMP considera "inaceitável e intolerável" não ser chamada a participar na elaboração dos regulamentos do Portugal 2020

[67]PEC—Plano de Estabilidade e Crescimento.

ANMP (2014b) Comunicado: Reforma da Fiscalidade Verde aprovada hoje na AR Governo e Assembleia da República não ouviram ANMP sobre proposta de lei. 5 Dec 2014
ANMP (2014c) Comunicado sobre as propostas de Lei que estabelecem o regime jurídico dos serviços municipais de abastecimento público de água, de saneamento de águas residuais urbanas e de gestão de resíduos urbanos. Coimbra. 5 Feb 2014
ANMP (2014d) Comunicado Acordo ANMP—Governo sobre o Fundo de Apoio Municipal. Coimbra. 8 July 2014
ANMP (2014e) Comunicado da ANMP sobre o Decreto-Lei no. 45/2014, de 20 de Março relativo ao sistema de gestão e tratamento de resíduos e autonomização deste sector no seio do Grupo Águas de Portugal. 20 March 2014
ANMP (2014f) Nota do Conselho Diretivo Proposta de Diploma: Estatuto das Estradas da Rede Rodoviária Nacional. Coimbra
ANMP (2015) Resolução do XXII Congresso da ANMP. Tróia, Grândola
Assembleia da República (2005) Constituição da República Portuguesa, 7ª revisão. Lisbon: Assembleia da República (english edition: http://www.fr.parlamento.pt/Legislation/CRP.pdf)
Barreto A (ed) (1998) Regionalização. Sim ou Não. Publicações D. Quixote, Lisbon
Bloco de Esquerda (2015) Manifesto eleitoral: Bloco de Esquerda, Legislativas 2015
Canotilho JG, Moreira V (1993) Constituição da República Portuguesa Anotada, 3rd edn. Almedina, Coimbra
Carita R (2007) 30 Anos de Autonomia, 1976–2006. Assembleia Legislativa da Região Autónoma da Madeira, Funchal
Carneiro JL (2012) A proposta de reforma da administração local: "O estado do debate". Direito Regional e Local 17:30–38
Caupers J (2009) Divisão administrativa e órgãos regionais. Direito Regional e Local 8:3–8
Coelho MH, Magalhães JR (1986) O poder concelhio: das origens às cortes constituintes. Centro de Estudos e Formação Autárquica, Coimbra
Coligação Portugal à Frente (2015) Agora Portugal Pode Mais. Programa Eleitoral
DGAL (2012) Estudo-piloto. Comunidades Intermunicipais Modelos de competências, de financiamento, de governação, de gestão e de transferência de recursos. Direção-Geral das Autarquias Locais, Lisboa
EC/ECB/IMF (2011) Portugal: memorandum of understanding on specific economic policy conditionality, 3 May 2011
Fonseca I (2013) Transferência (contratualizada) de atribuições e competências a favor das freguesias: possibilidade de ganhos e perdas? Direito Regional e Local 21:11–19
Gomes N (2015) O controlo da despesa pública no período pós-Troika: a versão 2.0 da Lei dos Compromissos e dos Pagamentos em Atraso. Questões Atuais de Direito Local 7:37–51
Gonçalves PC (2014) As entidades intermunicipais: em especial, as comunidades intermunicipais. Questões Atuais de Direito Local 1:21–40
Governo de Portugal (2011) Documento Verde da Reforma da Administração Local. Uma Reforma de Gestão, uma Reforma de Território e uma Reforma Política, vol 2. Lisbon: Governo de Portugal, Gabinete do Ministro Adjunto e dos Assuntos Parlamentares
Governo de Portugal (2015) Programa do XXI Governo Constitucional, 2015–2019
Lopes L (2009) O regime das Comunidades Intermunicipais: mais um caso exemplar de degradação da autonomia municipal. Revista de Direito Público e Regulação 2:9–18
MEPAT (1998) Descentralização, Regionalização e Reforma Democrática do Estado. Lisbon: Comissão de Apoio à Reestruturação do Equipamento e da Administração do Território
Miranda J (2007) Constituição portuguesa anotada. Coimbra Editora, Coimbra
Moreira V (2007) Associações intermunicipais e áreas metropolitanas. Direito Regional e Local 5–19
Moreno HB (1986) Os municípios portugueses nos séculos XIII a XVI. Editorial Presença, Lisbon
Oliveira AC (2008) 40 anos de desconcentração territorial regional em Portugal. Direito Regional e Local 1:5–13
Oliveira António Cândido (2011) É necessária uma reforma territorial das freguesias? Direito Regional e Local 13:5–9

Oliveira C (ed) (1996a) História dos Municípios e do Poder Local. Dos finais da Idade Média à União Europeia. Círculo de Leitores, Lisbon
Oliveira LV (1996b) A Regionalização. Edições Asa, Porto
Partido Comunista Português (2015) Programa Eleitoral do Partido Comunista: Legislativas 2015
Partido Socialista (2015) Agenda para a Década, 2015–2024. Moção ao XX Congresso Nacional. Proposta de Revisão dos Estatutos. António Costa—Mobilizar Portugal
PCM (2005) Programa do XVII Governo Constitucional 2005–2009. Presidência do Conselho de Ministros, Lisboa
Relvas M, Júlio P (2015) O outro lado da governação. A reforma da administração local. Porto Editora, Porto
Rocha JF, Pinto AM (2014) As finanças locais portuguesas após o 25 de Abril de 1974. Questões Atuais de Direito Local 2:43–67
Sá L (1989) Regiões Administrativas. O poder local que falta. Editorial Caminho, Lisbon
Santos JA (1985) Regionalização. Processo Histórico. Livros Horizonte, Lisbon
Santos JA (1995) As freguesias. História e actualidade. Celta, Oeiras
Silva CN, Syrett S (2005) Metropolitan Governance and Regionalism: the Case of Lisbon. In: Sagan I, Halkier H (eds) Regionalism Contested. Institution, Society and Territorial Governance. Ashgate, Aldershot, pp 247–264
Silva CN, Syrett S (2006) Governing Lisbon: evolving forms of city governance. Int J Urban Reg Res 30(1):98–119
Silva CN (1993) Réformes concernant la coopération inter-municipale et les limites: nouvelles tendences au Portugal. Bull Soc Languedoc Geogr 1993(3–4):281–299
Silva CN (1995) Poder Local e Território. Análise geográfica das políticas municipais, 1974–1994. Dissertação de doutoramento em Geografia, Universidade de Lisboa
Silva CN (1996) O financiamento dos municípios. In: Oliveira C (ed) História dos Municípios e do Poder Local. Dos finais da Idade Média à União Europeia. Círculo de Leitores, Lisbon, pp 433–462
Silva CN (2000a) A 'Região' em Portugal: do reforço da descentralização à fragmentação do Estado. Treballs de la Societat Catalana de Geografia, Barcelona XV(49):231–250
Silva CN (2000b) Regiões. In: Barreto A, Mónica MF (eds) Dicionário de História de Portugal, vol 9. Livraria Figueirinhas, Porto, pp 227–230
Silva CN (2000c) Local government growth and retrenchment in Portugal: politicization, neo-liberalism and new forms of governance. In: Hoggart K, Clark TN (eds) Citizen responsive government. JAI Press, New York, pp 223–241
Silva CN (2000d) Empresas municipais: um exemplo de inovação na gestão urbana. Cadernos Municipais-Revista de Acção Regional e Local XIV(69/70):13–20
Silva CN (2002a) Gestão de Áreas Metropolitanas: o "Modelo GAM-ComUrb". Cadernos Municipais-Revista de Acção Regional e Local XVI(81):42–63
Silva CN (2002b) Local and regional government: continuity and innovation in local governance. In: Syrett S (ed) Contemporary Portugal. Dimensions of economic and political change. Ashgate, Aldershot, pp 197–220
Silva CN (2004a) Portugal: Sistema de Govern Local. Barcelona: Institut de Ciències Politiques i Socials, Universitat Autónoma de Barcelona and Diputació de Barcelona
Silva CN (2004b) Public-Private Partnerships and Urban Governance: towards a new concept of local government? In: Barlow M, Wastl-Walter D (eds) New challenges in local and regional administration. Ashgate, Aldershot, pp 73–84
Silva CN (2006) Decentralisation and regional governance in Portugal: self-government or inter-municipal co-operation? In: Salet W (ed) Synergy in urban networks? Sdu Publishers, The Hague, pp 204–226
Silva CN (2009) Local political leadership in Portugal: excepcionalism or convergence towards a 'mayoral model'? Lex Localis J Local Self Gov 7(3):243–256
Silva CN (2014) The economic adjustment program impact on local government reform in Portugal. In: Silva CN, Bucek J (eds) Fiscal austerity and innovation in local governance in Europe. Ashgate, Farnham

Silva CN (2015a) Empowerment of local democracy and decentralisation of service delivery in local government reform: evidence from Portugal. In: Buček J, Ryder A (eds) Governance in transition. Springer, Dordrecht

Silva CN (2015b) Political and administrative decentralization in Portugal: a brief outline of the regional and local self-government system. Unpublished research note written for the Assembly of European Regions

Silva ST, Santos MC (2014) O Fundo de Apoio Municipal: algumas considerações. Questões Atuais de Direito Local 4:33–52

Syrett S, Silva CN (2001) Regional development agencies in Portugal: recent development and future challenges. Reg Stud 35(2):174–180

Chapter 3
Why and When Countries Implement Local Public Administration Reforms: A Long-Term View of Reform Dynamics in Slovakia, 1990–2015

Ján Buček

Abstract Public administration reforms are among the most typical expressions of societal progress from the point of view of governments and the public sector and more often than not have impacts on the institutional frameworks of urban governance. Debates and considerations on reforms are almost permanent, but their conversion into real terms and implementation requires much stronger motives. An overview of development in Slovakia from a longer term perspective shows that local public administration reforms usually need more complex stimuli. A combination of less positive indicators of social and economic development (induced by post-Communist transformation, or economic and financial crisis), need for progress in the field of local public administration and local development (democratization, decentralization, modernization), as well as the capacities of central governments and leading political parties, plays a primary role. Reforms have a better chance if there are stable (more electoral terms in central government) and well-established political elites. On the other hand, there are also factors that were influential only in a particular period and later on their impact decreased, or their nature changed. This is, for example, the case with administrative traditions and policy legacies. The modernization argument for reform has also changed—now strongly moved in favour of its technological and managerial meaning in Slovakia. International influences lost their strength since the time of pre-accession processes, although many piecemeal policy transfers are permanent. Surprisingly less reform calls and practical recommendations have been generated from within local public administration. Its elite is less compact and cohesive in generating larger scale proposals, although they are quite efficient in protecting previous reform achievements and in minor improvements to the public administration system.

Keywords Public administration · Reform · Local self-government · Implementation factors · Urban governance · Long-term view · Slovakia

J. Buček (✉)
Department of Human Geography and Demography, Comenius University, Bratislava, Slovakia
e-mail: bucek@fns.uniba.sk

© Springer International Publishing Switzerland 2017

C. Nunes Silva and J. Buček (eds.), *Local Government and Urban Governance in Europe*, The Urban Book Series, DOI 10.1007/978-3-319-43979-2_3

3.1 Introduction

Slovak local public administration has undergone many reforms and has adapted to numerous minor changes within the last 25 years since 1990. We can observe the need to build a new non-Communist local public administration, which emerged immediately after the Czechoslovak "Velvet Revolution" in 1989. This was later replaced by a different but still very challenging situation for the building of a new state after 1993, including the search for a new method of organizing local public administration. Transition and transformation process complications, combined with ambitions to join the EU in the near-future, generated pressure for a new wave of reforms at the turn of the millennium. Stabilization after joining the EU (in 2004) and a positive phase of the economic cycle were replaced by the "crisis years" leading to a reconsideration of local public administration functioning once again under the pressure of related fiscal consolidation.

Long-lasting reform efforts substantially changed the nature of local public administration in Slovakia, compared to the period before 1989, as well as the role it can have in urban governance processes. Manifold reform processes have included many processes such as decentralization, modernization, territorial reorganization, not mentioning modifications in public services delivery. These have had the most visible outcomes—the introduction of local and regional self-government that substantially influenced the nature of the country's government. They were part of the complicated transition processes across all sectors of society. While earlier stages of local public administration reforms in Slovakia have been quite extensively covered by scientific literature, this is not the case with the latest reforms. The attempts at longer term and more complex evaluations are also missing. Less attention has been paid to explain the timing of reforms, as well as a deeper consideration concerning factors supporting or limiting reform implementation in the field of local public administration.

Local public administration reforms within one country and in the long-term perspective are quite a frequent theme of scientific inquiry. Good cases represent, e.g., Wollmann's (2000) study on administrative reforms in Germany, Capano's (2003) study on reforms during the 1990s in Italy, or Alba and Navarro's study (2011) on administrative traditions and reforms in Spain. An interesting case focusing on Greek reforms within the context of the economic crisis was offered by Ladi (2014). Many reform-related issues covering a European comparative framework were done also by Kaczmarek (2005). In Slovakia, such longer term evaluations are missing. Nevertheless, particular stages of public administration reforms in Slovakia have been exposed by numerous studies prepared either by Slovak scientists (e.g. Buček 1993, 2006; Slavík 1997; Nižňanský 2002; Búšik 2005; Slavík et al. 2013), or by foreign scholars (e.g. Bryson 2008; Demmou and Price 2015). A more general context of reforms implemented during post-socialist periods in Slovakia, including an EU context, has been presented by, e.g., Mathernová and Renčko (2006), O'Dwyer and Kovalčík (2007), Bouckaert et al. (2011) or Buček (2012).

The main goal of this chapter is the search for relationships among local public administration reforms and the set of factors influencing their implementation and timing. Why and when countries prepare and implement public administration reforms is a considerable issue. Among the factors influencing reforms we must consider the role of administrative traditions and political legacies. Reform dynamics also reflect the international situation and integration ambitions. Modernization is an important part of reform motivation. More detailed attention is paid to the conditionality of local public administration reforms on macroeconomic development and the public finance situation. Within the political framework, we focus on the role of central government, its political composition and duration in holding power, as well as political priorities and the role of political parties. We also perceive reforms as governance-based output, with many actors involved. Of course, dealing with reform experience in one country means particular limits in generalization. Nevertheless, it provides some interesting knowledge for countries considering reforms, e.g., during more turbulent and transitional periods of their societal development, or offer a knowledge base suitable for comparative studies. It also contributes to the understanding of reform approaches under the dual model of local public administration (with separate lines of state administration and self-government adopted in Slovakia since 1990).

In Slovak case, we focus on local public administration, integrating the separated lines of state administration (operating on various sub-state levels) and self-government (local and regional self-government). Nevertheless, with respect to the diminishing role of state administration over time, more attention is paid to the institutions of territorial self-government (we primarily focus on local self-government due to its longer functioning). As a general framework, we take into consideration Central East European experiences with public administration reforms; however, within the framework of global trends, public administration has been facing over the few last decades. In this study, we do not pay broader attention to such aspects of reforms as civil service reforms, or reforms in central public administration (e.g. central government ministries, or specialized state agencies). The main sources of information include important public administration reform documents and legislature, supplemented by documents and comments prepared within the legislation process. We also use available studies, as well as statements by the important actors (e.g. representatives of particular governments, or important associations). The reliable long-term macroeconomic and public finance data needed for the evaluation were taken from the databases of OECD, the Slovak Statistical Office and the Slovak Central Bank—National Bank of Slovakia.

3.2 Conceptual Background: Factors Influencing the Reasons and Timing of Reforms

Local public administration reforms are prepared and implemented within a wider framework of other spheres of social life (political, economic, cultural). The understanding of driving forces and their temporal circumstances can contribute to revealing why and when reforms have more chance of being adopted and successfully implemented. Such knowledge can be achieved by applying more interconnected concepts. As a key approach, we were inspired by historical institutionalism, including the impact of critical junctures, frequently applied in the study of long-term changes in societies (e.g. Capoccia and Kelemen 2007; Steinmo 2008; Hall 2010; Mahoney and Thelen 2010). We focus on institutional changes perceived as changes in rules and organizational framework concerning the sub-state level of public administration. Complementary concepts taken into account are those relevant to the study of key reform processes such as democratization, self-government introduction (local and regional) and decentralization as general tendencies in managing public affairs in Slovakia after 1989. Transition studies are also a useful conceptual framework, in which the establishment of self-government and its development is considered as an important segment of the overall transition from Communism. Within this transitional framework, important roles were played by reform theory studies. Several authors have written about post-Communist reforms, stressing the importance of various sections of reforms (e.g. Fidrmuc 2003; Mathernová and Renčko 2006; Myant and Drahokoupil 2013), or paying more attention to public administration reforms (e.g. Péteri and Zentai 2002; Verheijen 2007). From interpretation point of view we also were inspired by coincidence method (e.g. Bothe 1955), focusing on linkaged among simultaneous event-related conditions.

Public administration reforms, including those focusing on the local level, are important societal innovations that are the frequent subject of scientific inquiry. Among the key starting points, we have to explain is: What kind of changes in local public administration institutional framework can be considered as reforms? It is usually the case of large-scale changes in local public administration, accompanied by a cluster of new (or amended) legislation, in some cases including even constitutional changes. Reforms contain at their core, e.g. the launching of a new level of political organization, changes in the scope of local autonomy, transfer (redistribution) of important powers (administration, services), extensive changes in financial flows and related rights (e.g. in the field of taxation, budgeting) and the transfer of property (e.g. linked to particular powers). Major reforms often intervene in spatial issues, primarily in the form of spatial reorganization (e.g. in territorial administrative division, amalgamation), or the relocation of offices. We can observe more "strategic" and complex reforms that substantially reorganize the public administration system and initiate major changes, as well as more narrow-oriented managerial and technical reforms. Selected cases of public administration reforms classifications have been summarized by, e.g., Kaczmarek (2005). Countries usually oscillate between periods of major reforms (which are rare in well working public administration systems) and periods typified by less extensive changes, following the practice of incremental changes to an already existing institutional framework (spread over many years). Such minor "piecemeal" and isolated modifications to an existing system we do not consider as reforms. We take into account the scope of changes and not if they have been formally declared as reforms (e.g. by the government elite).

Any attempt at the evaluation of longer term reforms requires more attention paid to time aspects, closely related to the preparation, management and implementation of reforms. The dynamics of reforms influence also the attitudes of leading political and social actors, those working in public administration, available expertise and public perception. It is not easy to characterize precisely a particular period as a radical/shock reform period, a period of gradual/incremental reform, or a stable period without any significant changes (no reform). We have to be aware that public administration reforms (depending on the scope of reform), often concern many years, have more stages and represent more demanding cases of policy-making processes. Almost each reform has its preparatory stage typified by analytical works, strategies elaboration, search for political support, preparation of executive and legal documents and so on. Reforms have their most visible stage during the implementation period, including the solution to important practical issues (implementation management, its organization, financing, staff and so on). Each reform change leads to subsequent evaluation after introduction into practice. Reforms need to be assessed after a certain amount of time, as well as often inevitable post-reform adaptations based on post-reform practical functioning experience. So it is a matter of careful consideration how to take into account all stages, or focus on a particular stage. Reforms are usually evaluated as a unity of more phases, although with a larger emphasis placed on period of their factual implementation.

The decision to make any public administration reform adoption and implementation depends on the various factors and their role in particular circumstances, including the rising impact of exogenous context. The impact of such factors motivating reforms is diverse, and some of them are also internally structured. Some factors influence the nature of the reforms, while some also strongly influence their timing. Reflecting on international experience and taking into account the factors usually mentioned (e.g. Wollmann 2000, 2012; Illner 2003b; Baldersheim 2014) as those standing behind changes in local public administration, we are focusing on the role of:

- administrative traditions and policy legacies,
- modernization,
- international political environment and integration processes,
- macroeconomic development,

- public finance development,
- central government and policy priorities,
- political parties,
- governance and other actors.

Local public administration reforms usually have to take on the administrative traditions and policy legacies of particular states (e.g. Meyer-Sahling 2009). Some of these are deeply rooted within the society and have their strong advocates. Among the basic settings influencing reforms, we have to mention the historically developed settlement system (e.g. its fragmentation, Swianiewicz 2010), population distribution characteristics, territorial division and institutional traditions. In CEE countries, among the important determinants we can mention memories of the practice of public administration from the interwar period (e.g. in Czechoslovakia), or even traditions based on the previous centuries of local government functioning (e.g. from the long-lasting Austro-Hungarian administration). There have been great expectations based on the simple transfer of historical experiences, often fragmented, without detailed knowledge and less suitable for practical implementation in the present. Such feelings had been expressed enthusiastically concerning the "return" of older "rights" to the local level after 1989 in many CEE countries.

Long-term approaches cannot avoid thinking about historical influences within the path dependency framework (e.g. Pierson 2000). Within the post-Communist framework, we cannot overlook the influence of public administration practices applied during communist period that had not been so easy to overcome (e.g. Illner 2003a). Such path dependency cannot be underestimated at least in some post-communist countries. Waiting for commands for and solutions to local problems "from above" had been quite a frequent approach in the early years of local self-government functioning. Traditions, the Communist legacy and their proponents had an important impact on the willingness for reform, its framework, phasing and the dynamics of implementation. Another issue is to what extent new decisions have generated new path dependencies and have influenced the next stages of reforms. They can mean a burden that limits progress in particular fields of government.

It is usually accepted that in many cases reforms are driven and modified by the relevant international political framework in general, as well as by specific policy transfers (Dolowitz and Marsh 2000). These were the experience of Western European countries that were applied in Central Eastern European public administration reforms (e.g. Baldersheim 2014). One very important aspect was the effort to harmonize public administration system with models and principles working in Western European countries. This feature was particularly important during the effort to join the EU. "Europeanization" had a big impact on changes in public administration throughout the CEE (Grabbe 2001; O'Dwyer 2006; Kovács 2009). The ambition to satisfy expectations, conditions and standards of the EU mobilized activity in reform preparation and implementation. It was typical, especially during the second half of the 1990s and at the beginning of the next decade, in many CEE countries, sometimes referred to as the "pre-accession wave" of reforms (Bouckaert et al. 2011). It is also important to mention large-scale policy transfers, inspiration or even policy imitations concerning local government that reflected the practice in well-developed local governments in Western countries. The nature and speed of reforms in many CEE countries were also related to the leading role of neo-liberal approaches within policy transfers at the end of the twentieth century (and the weaker role of other alternatives).

Modernization, in various meanings, is among the usual drivers and reasons for changes in public administration. In its very general sense, modernization as a progress can be considered a permanent factor of change, when "new and better replaces the old". It has been frequently used as one of the key arguments within main reforms, or it has been used to describe less extensive innovations in public administration. Many reforms in developed countries had been implemented as modernization reforms, with the ambition to provide a better public service (e.g. Connaughton 2008 on Ireland). Beside a possible extensive philosophical debate on modernization in public administration (e.g. Illner 2003a, b), we can apply a more pragmatic approach. Modernization is mostly perceived in its political (democratization), administrative/managerial and technological concept. Undoubtedly, the public administration system in CEE countries needed manifold modernization after the change of regime. By far the most urgent was the democratization that started immediately after the change of regime in most CEE countries. Modernization, from a more narrow view, has usually focused on administrative and managerial innovations—dealing, e.g., with efficiency, accountability, better services provision, transparency and staff training. For a long time, modernization in public administration was also strongly linked to the concept of new public management (e.g. Wollmann 2012). Public procurement, contracting out, privatization of previously public services, public-private partnership and customer orientation had become more frequent, although applied later and in a fragmented way also in CEE countries (e.g. Bouckaert et al. 2011).

Within the last two decades, a more challenging aspect of modernization has been the application of information and communication technologies (ICT), as a shift towards digitization, e-government and technologically based participation and communication with citizens, e.g. in local development planning (e.g. Brown 2005; Dunleavy et al. 2006; Silva 2010). This technological meaning of modernization is very influential in current scientific discourse as well as in practice concerning local public administration. It is often represented by a series of incremental changes and innovations implemented at all levels of public administration, with many innovations generated by individual local governments, or induced by the available technological advancement. The governmental sector has the ambition to be compatible also technologically with development in other parts of society, which is reflected in public administration reforms.

Risk and benefits related to public administration reform implementation have to be evaluated from a macroeconomic perspective. Local public administration reforms as important policy decisions are adopted in a particular macroeconomic situation, related to major macroeconomic policies, as well as social and economic development at regional and local levels. We have to take into account such relations, although they are not so simple or easy to identify (Péteri and Zentai 2002). The economic context of public administration reforms cannot be ignored. Political cycle literature confirms the serious concerns of politicians/political parties about economic performance and extensive exploitation of macroeconomic policy (e.g. Hibbs 1977; Potrafke 2012). Reforms can touch an important section of the economy under the direct control of public bodies, or be regulated by the central state. Good social and economic conditions can lead to a "no reforms" approach, or on the contrary, are suitable for reforms whose implementation require, for example, higher costs and a wider acceptance of reforms. A bad macroeconomic situation can also initiate debates on inevitable complex changes that can also extensively influence the local public administration system. A typical case is the situation during economic crises, when governments intervene extensively not only into the economy, but also into public policies and public administration. The regional and local social and economic situation can also have a diverse impact. A worse situation in many regions can multiply calls for reforms. New institutional arrangements can improve their prospects for future economic growth. On the other hand, any attempts at reforming can be evaluated according to their impact on the regional and local economic and social situation. The need for other reforms (e.g. economic) can lead to postponement or the inclusion of public administration reforms into a wider package of reforms. In the CEE situation, inevitable post-socialist transition reforms could influence the timing of public administration reforms. Among the usual indicators of a social and economic situation taken into consideration, we can find GDP growth, unemployment rate, inflation and interest rates.

Strongly related to the macroeconomic reasons for reforms is public finance development. A worse macroeconomic situation (e.g. caused by an economic crisis) and a too unbalanced budgetary policy (e.g. public budgets deficits accumulating) can have a devastating effect on public finance. The need to solve a bad public finance situation can lead to various reform measures focusing on public finance savings, including those addressing local public administration. This is true that despite an already widespread application of fiscal decentralization and a certain level of local fiscal autonomy. Besides quite often contradictory studies analysing local governmental austerity policies in crisis situations (e.g. Bakota 2014; Buček and Sopkuliak 2014; Ladi 2014; Silva 2014), a suitable knowledge framework for debating the impact of public finance on reform efforts is provided by consolidation studies. They analyse experiences with public finance consolidation in many countries, often in the long-term and with a comparative perspective (e.g. Blöchliger et al. 2012; Dellepiane-Avellaneda and Hardiman 2015). According to Perotti (1998), fiscal consolidation has the features of policy reform because it is, for example, reforming the budget process, it might involve changing the government's employment policies and the structure of public transfer programs. Typical for consolidation periods are measures focusing on public budgets and public debt reduction. Under such a regime of financial scarcity, measures can address not only central state institutions, but can induce important changes for the sub-state governments as well (see e.g. Rodden and Wibbels 2010). Local budgets are an important part of public budgets in general. A successful consolidation effort requires a coordinated approach across the entire public finance sphere, including fiscal consolidation at lower levels of government (e.g. Molnar 2012). Nevertheless, all reformers must be aware that reforms also have their own costs. Part of the costs can later return via less costly functioning in the future. The usual indicators studied under such a situation are public budget deficits, public debt (e.g. towards GDP), including local budget deficits and local public debt.

Central government composition, its internal stability, duration and leadership efficiency affect progress in public administration reform. The key justification of its position is related to the usual centralization of an administrative reform effort and its dependency on central state leadership (see e.g. Aucoin 1990). Political environment development and electoral preferences are important for the formation and composition of central governments and their approaches to reforms. Preparation and implementation of reforms can be easier under a single party central government, if they are really interested in reforms. However, in Central Eastern European countries, single party central governments are quite rare (e.g. Müller-Rommel et al. 2004; Conrad and Golder 2010). The prevailing electoral results have led to coalition central governments responsible also for initiation and implementation of local public administration reforms. This aspect of central government party composition can play an important role in the scope, speed and efficiency of reform activity. Too many parties and/or too diverse governmental coalitions can influence reforms in general by diverse or even contrasting approaches and can lead to less extensive or compromising reform outcomes, or obstacles in implementation. A government composed of similar parties can more efficiently prepare and implement reforms. Diverse and less clearly identifiable political parties influence another aspect of reform potential, which is government stability and duration. Stable coalitions and a full electoral term or even better—a stable central government holding power for several terms, can support the completion of reforms. A re-elected reform government can complete all the details of reform objectives, including eventual corrections. Among other factors we can mention leadership of a government coalition and their interest in reform. The leader of a coalition must have an interest in implementing reforms (including the influence of the prime minister, or other strong cabinet member of). Another part of the central government is the central state bureaucratic elite, which is often less interested in many aspects of reform, including decentralization (less powers and resources under its direct control).

Political parties and their interests have an important impact on administrative reforms. The reform initiation and implementation responsibility is in the hands of political parties and structures of government under their control. Their position is predisposed also by the electoral system and electoral cycle, the political parties' system and its stability. Political parties' opinions on the nature, scope or timing of reforms are crucial. Nevertheless, their opinions have been diverse, depending not only on the traditional right-wing and left-wing divide. Major debates can concern, e.g., their opinion on the scope of centralization and decentralization, or the position of particular levels of government, or scope of regulation (e.g. in Péteri and Zentai 2002; Illner 2003b, Klimovský 2008). A fragmented political system with many new political parties, with the absence of solid and stable support, and an unclear ideological background (left/right divide is less useful at least in some countries for particular period), can complicate progress in reform processes. This is especially the case in countries with "young" political parties and less balanced political systems missing deep democratic political traditions, especially during the early phases after a change of regime. Less stable and internally less cohesive parties can have a less clear position concerning local public administration organization, or opinions on necessary change. It can influence the search for the stable support of reforms or threaten their implementation. The willingness of political parties to support reforms can be influenced by their ability to achieve a significant position within particular levels of government (e.g. depending on the situation in their local party structures). It has also been mentioned that for the post-socialist situation, it is the centre that has better control, with an unwillingness to lose control at the local level by the dominant parties. The opinions of key political leaders can also influence the direction of reforms, as well as the efficiency of their implementation. The direct support and involvement of leaders can have a positive impact on their final implementation.

Democratization after 1989 offered freedom of activity to many new societal actors. This trend was supported by penetration of governance as a new principle in public policies adoption (e.g. Peters and Pierre 1998). Under new conditions, any substantial change in institutional settings under a democratic regime requires an agreement among the important societal actors (e.g. Hall 2010). Administrative reform is in fact a wide-scale co-ordinated collective action. Besides the dominant role of political parties, or representatives of large segments of society, those most directly influenced by changes in the local public administration system have an increased role. It is also multi-level governance activity. Much larger roles are held by the rising strata of local and regional leaders, as well as their representative associations. They represent an important segment of society, disclose opinions from regions, reflect their social economic conditions, administrative capacity and general willingness to implement reforms. We also cannot underestimate the role of bureaucracy at all levels that can also be influenced by reforms. Their acceptance or rejection of reforms can be important in the speed of their implementation. It is especially important if we think about reforms with objectives in decentralization, reorganization and deeper modernization (e.g. also with a reduction in the number of staff).

3.3 The Slovak Local Public Administration System and Its Reforms in Brief

The main feature of the Slovak local public administration system is its dual nature consisting of separated lines of state administration and self-government (not answerable to one another). This model, adopted in 1990, has influenced all reform intervention into the public administration system. Both lines underwent more reforms during the post-socialist period after 1989. They could be parallel, as well as separate. They could have different timing, as well as priorities. The reform periods in Table 3.1 outline the years with major changes, including major conceptual documented elaboration, as well as reforms implementation. It documents the concentration of major reforms at the beginning of the transition period, at the turn of millennium and prior to joining the EU and after the financial and economic crisis and its transposition into the public sector.

Long-term separate development under a dual model of public administration has had consequences for their diverse organization. Currently, there are two levels of self-government under the central state. The lower level known as local self-government (in Slovak—miestna samospráva) was established in 1990 and is now composed of 2891 units (average size is below 1900 inhabitants). The upper level—eight regional self-governments (in Slovak—regionálna samospráva), was introduced in 2002 (the first regional elections were held in 2001). The situation within the sub-central state administration was more complicated with more levels and spatial units used after 1989 (e.g. "old" larger and "new" smaller districts, higher territorial units with a regional dimension, or specific regions with respective offices of specialized state administration). Sub-central state administration is currently organized primarily at the district level with 72 district offices (in 79 statistical LAU1 districts in Slovakia, including urban districts in Bratislava and Košice).

3.3.1 Local Self-government Reforms

Renewal of local self-government on the basis of already existing local administrative units was one of the key features of the changes after 1989. It reflected calls for immediate democratic changes at the local level, based on the tradition of self-government, e.g. from the interwar period. Its role had been increasing thanks to more phases of major reforms, as well as from frequent minor adjustments after 1989. The major reforms were typified by more extensive changes—with important legislation changes, including changes in the distribution of power and resources. Among the major reforms, we can observe numerous incremental changes (technical and managerial changes, clarifications and improvements to previously adopted legislation). Besides the reforms that were implemented, there also were reforms which were not adopted. For example, this was the case with so-called communal reform (in Slovak—komunálna reforma), addressing the problem of the large number of small local self-government units in Slovakia, or related attempts for more efficient provision of important powers, such as territorial planning and urban development. Even the adopted reforms were the subject of disputes and had been modified before final adoption in the Slovak Parliament.

The first cluster of major reform changes in local self-government we can characterize as early transition reforms in 1990–1991. This was focused on

Table 3.1 Main periods of local public administration reforms—schematic outline of years with reform

	1990	1991	1992	1993	1994	1995	1996	1997	1998	1999	2000	2001	2002	2003	2004	2005	2006	2007	2008	2009	2010	2011	2012	2013	2014	2015
Self-government	x	x								x	x	x	x	x	x	x				x	x	x	x	x		
State administration	x	x	x	x		x	x			x	x	x	x	x	x			x					x	x	x	x

Source Own interpretation

breaking up old administrative structures, introducing democratic local self-government and its institutions, allocating selected fundamental powers, transferring property, as well as formulating the basic administrative and regulatory framework for local self-government (e.g. Acts No. 369/1990, No. 517/1990, No. 138/1991 Coll). Besides this, abolition of the old regional level of government (four regions, in Slovak—kraje, krajské národné výbory) was also very important. In fact, the extent of self-government had been limited. The scope of powers had been limited, when most power was in the hands of the state administration. They also suffered a worse financial situation, with little real independence as far as resources were concerned (e.g. Bryson 2008). During this introductory period, local self-governments paid attention to the building of their democratic functions, formation of their administrative capacities, seeking possible ways to manage allocated powers and consolidating their own property. State-building processes and other transition processes did not allow their strengthening after the first years following the change of regime.

The second important reform concerning local self-government was put into practice during the late-transition period in 1999–2005. It was a longer reform period due to the fact that it was the key decentralization reform, which also included extensive conceptual preparatory activities. The central government adopted two strategic documents (Strategy of public administration reform in the Slovak Republic, 1999; Concept of decentralization and modernization of public administration reform in the Slovak Republic, 2000). The intentions behind the reform had been extensively debated with local self-government representatives and the public (e.g. Nižňanský 2002). This quite complex reform included the introduction of regional self-government (Act No. 302/2001 Coll.), the sequential transfer of a large number of powers and property from state administration to self-government, respecting the absorption capacities of local self-governments. Another large-scale change was the financing of local self-government, thanks to fiscal decentralization introduced into practice in 2005. This reform substantially strengthened the role and resources administered by local self-government. Nevertheless, as Buček (2006) outlined, the decentralization was not a one-directional and unconstrained process. There remained important delegated and shared powers with a large state administration involvement. New powers or resources were balanced by rules not applied previously (e.g. stricter budgetary rules, restrictions on borrowing).

The third cluster of changes in local self-government can be considered as being unplanned (without clear "reform document"), permanently negotiated, less complex, less extensive and a more-or-less forced "counter-reform". This was induced above all by the global financial and economic crisis and its later shift to the public sector. Reform period started by signing memorandum on cooperation among Slovak central government and Association of Towns and Communes of Slovakia in 2009 (see e.g. Buček and Sopkuliak 2014). Its different nature confirmed intervention into previously adopted central-local relations and the scope of local autonomy primarily in the field of local finance and budgeting (the most typical expression of this period is frequent changes in budgetary rules Act No. 583/2004,

Constitutional Act No. 493 adoption in 2011, Income Tax Act 595/2003 amended in 2011 and next memorandum signed in November 2012). Such development significantly influenced various aspects of local life and the activities of local self-governments. This post-crisis reform with main decisions adopted and changes implemented in 2009–2013 can be considered as "interim" and consolidation related, as it has been incorporated into the public finance consolidation efforts of the central government. It was reduced in its impact upon self-governments as the situation in public finance improved during 2014–2015. This period of development ceased further progress in favour of local and regional self-government strengthening and their activities in local and regional development. This period is considered as being specific, with less changes having a long-term character.

3.3.2 State Administration Reforms at the Local Level

The dual model of public administration has meant that besides more general administrative reforms, part of the reforms has concerned only local state administration. This line of public administration is easier to intervene in, being directly subordinated to the central government. It has resulted in more frequent changes, including those of a technical and managerial nature. In fact, local state administration development faced instability, changing approaches, an absence of a longer term vision, oscillations between separatist-sectoral and integrative approaches, not mentioning subordination to the political interests of the central government political parties. We identify four main phases of state administration until now—1990–1993, 1995–1996, 2002–2007 and 2012–2015.

The early transitional reform was parallel to local self-government, although it was longer (1990–1993), due to state administration adjustment into the new situation of the new state and its needs. The main inevitable changes were introduced after 1991. Local state administration had operated in 38 district offices and 121 area offices of state administration. The creation of an additional lower level of administration (area office) was soon considered as being useless. A specific feature of the state administration had been the formation of many networks of specialized state administration, operating mostly on the sectoral principle as field offices of particular ministries (e.g. labour offices, environmental offices, tax offices). This process was uncoordinated and inefficient (see e.g. Slavík et al. 2013). Particular networks were diverse in the horizontal, as well as vertical, dimension. Local state administration became very fragmented and complicated for citizens.

After the interim stabilization and some short-term preparation, a larger scale and separate state administration reform was introduced in 1996 (Acts No. 221 and No. 222/1996), without any widespread public discussion. This was a serious change in the organization of the whole state administration system, accompanied by the introduction of new district and regional state administration offices, with a new territorial administrative division of the country (79 districts and 8 regions). It also accompanied the partial integration of fragmented sectoral state administration offices (e.g. Búšik 2005). This reform can be also considered as part of the state-building processes of that period's central government. This reform lacked mutual agreement across the political spectrum and many aspects of it were disputed (e.g. the territorial administrative division of the country).

The next reform had been logically parallel and integrated with the self-government and mainly reflected the transfer of powers from state administration in favour of local and regional self-government during 2002–2004. Wide-scale decentralization led to the elimination of district offices of the general state administration in 2004 (Act No. 515/2003). They lost a substantial part of their previous powers. As a result, a network of new area offices was established, and certain networks of specialized state administration also came into existence. For similar reasons (after a successful takeover of power), the next reorganization of state administration cancelled the regional offices of general state administration in 2007 (Act No. 254/2007). Their powers were taken over by area offices in regional centres.

A new wave of changes was implemented under the influence of public finance consolidation in 2012–2015. The main legislature was adopted in 2012–2013 (Act 345/2012 and Act 180/2013), while implementation continued over the next few years. Among the main reasons for reform was the expected extensive savings in public administration costs (at the beginning this was estimated at EUR 700 million between 2012 and 2016, Pravda 2014). This reform was announced under the name "ESO Programme" (in English, stressing efficient, reliable and open state administration). It again focused on the lower level and a network of 72 integrated district offices was established. It integrated previously disbanded field offices of state administration based on the sectoral principle into one office (see e.g. OECD 2014). Regional offices of specialized state administration were also cancelled. It included technological modernization (more information technologies were introduced), with a more customer-oriented approach. This reform and its cost demanding segments (hardware, software for e-government, buildings' reconstructions) was possible to implement mostly thanks to the extensive use of EU funds. It resulted in opening a network of one-stop-shop-type client service centres covering all citizen agendas in one place.

3.4 Factors Influencing the Implementation of Local Public Administration Reforms in Slovakia, 1990–2015

3.4.1 Administrative Traditions and Policy Legacies

Traditions influenced public administration reforms in a diverse way in Slovakia. While on the one hand they allowed the quick application of models inspired by positive historical experiences, on the other hand discussion about "traditions" limited progress in some fields of public administration reforms. There are also signs of path dependency that influenced public administration reforms. Old legacies were still valid, e.g. in the case of the large number of small local self-governments, or territorial division of ethnically mixed territories. The transfer of staff from state administration to self-government had a similar impact, which limited the application of new approaches at least for a short while. It is also questionable, if the new territorial division at the regional level introduced in 1996 would develop into a new path dependency factor circumscribing progress at this level.

The existence of well-known traditions was important for the quick introduction of local self-government in 1990. The early stage of the reform had not been extensively planned. There was a lack of time for deeper scientific and political debates, so inspiration taken from previous eras was spontaneously adopted. There were also tradition-based proposals, e.g. concerning the regional level of government (e.g. territorial division), but such proposals were rejected as being outdated. Institutional traditions later faced a more consolidated political environment and a different phase of social and economic development with changing opinions on the possibilities to implement a historically justified approach. A historically reasonable approach was adopted, e.g., in the field of returning old municipal property back to local self-governments (property they had owned before the Communist period).

One kind of path dependency we can observe in the case of inherited administrative-territorial division at the local level typified by the large number of units. The number of local self-government units increased during the first years of the transition period as a part of the freedom that returned to the local level. Nevertheless, it limited progress in decentralization and also in the present it influences the efficiency of local self-government (administration, services delivery). Debates on the too fragmented network of local self-governments are on-going (e.g. Slavík et al. 2013; Klimovský 2014). The specific arrangement concerning the territorial administrative division of ethnically mixed territories in Southern Slovakia (e.g. as "petrification" of old districts) had a similar impact. Specific path dependency influenced the limited progress that public administration had with the simple transfer of staff from the previous state administration to self-government (e.g. during the early years of regional self-government). They took with them their previous practices of state administration officials.

As a new path dependency formation, we can consider decisions concerning spatial division at the meso-level adopted in 1996. Territorial division which does not respect any clear rational requirements, or traditions, with an absence of regional identity, is in a fragile position. It faces less respect among political strata, as well as among citizens (see e.g. Buček 2011). It leads to a weaker position of regions and their representatives within society. It is a matter of future development if this path dependency burden will be satisfactorily resolved.

3.4.2 Modernization

Modernization has been a common argument in favour of reforms in Slovakia. It is not surprising that after decades of a Communist regime, it has been repeatedly declared as part of the general effort for inevitable progressive development. We can observe all the crucial meanings—democratization, decentralization, managerial and administrative modernization, as well as information-communication and technological modernization. Although they are intermingled, there are periods during which particular meanings prevailed in reforms' implementation.

There was a strong need for democratization felt immediately after the fall of the totalitarian regime. The collapse of communism also opened the opportunity to introduce influential neo-liberal principles of capitalism, including the preference for rapid "shock" reforms implementation. A strong emphasis on economic reforms (as privatization and a market economy were introduced) limited interest and capacities for parallel deeper administration reforms immediately after 1989, besides democratization. A more extensive meaning of democratization expanded later, within the next reform periods, focusing on, e.g., a more participatory government, better information access and transparency in public policy.

More explicit pressure for public administration modernization was explicitly included in the key 2002–2005 reforms. This had been indicated already in reform preparatory documents elaborated since 1998. Modernization had been one of the primary goals expressed within *Concept of decentralisation and modernisation of public administration* adopted by the central government (Office of the Government of the Slovak Republic 2000). It focused on managerial, administrative (civil service principles), staff training and ICT meanings of modernization.

Management and administration modernization focused primarily on improvement of efficiency, quality and transparency in local public administration (primarily in self-government). A more visible penetration of various new practices in administrative and managerial fields had been observable already since the mid-1990s. Nevertheless, it was multiplied after the key decentralization reform came into effect. It was a reaction to the demand to build a more efficient, better organized and initiative self-government. Partly, it had been inspired by the principles of new public management (well visible e.g. in ESO state administration reform). It concerned, e.g., performance measurement, programme budgeting, public procurement, customer orientation, public-private partnerships (e.g. Malíková et al. 2013). One specific issue had been the improvement of the quality of staff, its legal status, with pressure on better training and education of staff working at all levels of public administration.

Greater attention to the "technological" meaning of modernization within reforms (with many headlines changing over time, such as informatization, electronization, digitalization) also started after the year 2000 (they were progressing also before, but as uncoordinated activities, e.g., of individual self-governments, or separate lines of state administration). It focused on preparation of strategic and legal framework, pilot projects and fragmented applications. Later on, interest shifted to integration of fragmented segments of information infrastructure and already developed applications. The role of ICT in the improvement of decision-making and a better service for citizens has been generally recognized. Despite progress in this field, it has been considered as being less sufficient compared to in other countries. As a result, since 2008, the modernization effort in this field has increased and has shifted in fact into the core of reforms. A specific strategic and conceptual reform document focusing on informatization in public administration was adopted in 2008 and amended in 2015 (Národná koncepcia informatizácie verejnej správy, Ministry of Finance of the Slovak Republic 2008, 2015). Progress in this field led to improvement of e-government in Slovakia by a set of incremental improvements and also thanks to great support provided by EU funds (with delay to the end of programming period). It would be the technological aspects that would prevail in the reform effort if consolidation pressure did not press for other reform steps. Dominance of information technological aspects in reforms confirmed plans to establish the Council of the Government for Public Administration Digitalization (Slov. Rada vlády Slovenskej republiky pre digitalizáciu verejnej správy, Office of the Government of the Slovak Republic 2015). Besides representatives of public administration, it should include also representatives of the IT sector (e.g. associations active in this field).

3.4.3 International Policy Environment and Integration Processes

The international policy environment we can consider as a source of important external influence in general. This has been especially true in the case of smaller countries, with open economies and strong integration ambitions. An effort to have a standard local public administration system comparable with other EU or OECD countries was combined with the goal for dynamic social economic development in Slovakia. While policy and good practice transfers work almost permanently, EU integration ambitions strongly affected the timing of reforms and pressure for their implementation. Slightly different was the situation during the global financial and economic crisis. Although it was primarily an "imported" crisis (e.g. Buček 2012), with less internal sources of crisis in Slovakia, economic turbulences and rising public debt led to consolidation policy adoption, which influenced local public administration. Nevertheless, despite various international influences, we can still say that the local public administration system still has numerous country-specific features.

Public administration reforms in Slovakia were influenced by the experiences and traditions of selected Western countries at the very beginning after 1989. They concerned countries closer geographically, with a similar historical background, administrative traditions and good experience in public administration, with important influence of German–Austrian local government tradition. Of course, there has always been inspiration from reforms in other post-socialist transforming countries, primarily the Czech Republic (until 1992 in one state), as well as Hungary and Poland. The Slovak Republic carefully observed their reform activities in the field of public administration, with the goal of not staying too far behind them in these kinds of reforms (good local public administration is often perceived also as an important aspect of competitiveness). Later on, external environment impulses were related to the integration of Slovakia into the EU, OECD and NATO (end of the 1990s). The most influential role was "Europeization", typified by the European Union conditionality and recommendations prepared, e.g., by OECD. They were often included in argumentation within reform documents and legislation preparation in Slovakia, as well as in other countries (in the case of Portugal, e.g. Magone 2011). These integration ambitions also had real-time influence and sped up reform activities in the field of local public administration. Thanks to integration process-related policies and "harmonization", many innovations were introduced also in the field of local finance (e.g. programme budgeting), services provision, ICT application and so on.

Many important policy innovations transferred from abroad entered into the basic principles of the reforms. Among the most important external influences, we can consider the key ideological and conceptual shift. We can observe the penetration of influential concepts such as decentralization, subsidiarity, new public management, governance, not mentioning strong a neoliberal background (sometimes in fragments). They were transferred by various means. It included rising participation in international bodies, for example, in the Council of Europe. The Slovak Republic signed the European Charter of Local Self-government in 2000 (selected Articles) and in 2007 (in full extent). Many new policy innovations penetrated thanks to rising decentralized international cooperation (twinning, sister-cities, cross-border co-operation), especially into policy-making and services delivery practices of self-government. Such external "policy shopping" is permanent and plays an important complementary role in domestic approaches in shaping the final goals of public administration reforms.

3.4.4 Macroeconomic Development

The Slovak economy has experienced quite successful development during the last 25 years (Fig. 3.1). Nevertheless, this development has not been straightforward and interim turbulences have emerged as well. The relationships among macroeconomic development, economic policy and public administration reforms within the state have been important. Post-socialist countries such as Slovakia had been concentrated on reforming their fragile economies for many years. We have to be aware that public administration was not considered a priority within the main transition processes (e.g. Verheijen 2007). Market economy formation and related economic reforms dominated political discourse. Within the economic transition policy, preference had been given to "shock therapy". Inside this economic framework, public administration reforms had been considered as a secondary issue during the 1990s, although still important. At the turn of the century, the public administration reform position changed. It was integrated into mainstream economic and social reforms induced by significant economic stagnation of the country in this period (see Figs. 3.2, 3.3, 3.4). Public administration reform was once again included in the responses to an economic decrease during the economic and financial crisis and afterwards. It shows that public administration reforms are induced also by worse economic development in a country. Such a situation provides great opportunities to build a wider consensus on changes also in this field, under the pressure to adopt inevitable and deeper changes within society. The Slovak case shows that it makes sense to make reforms, because a better phase of economic and social development follows their implementation.

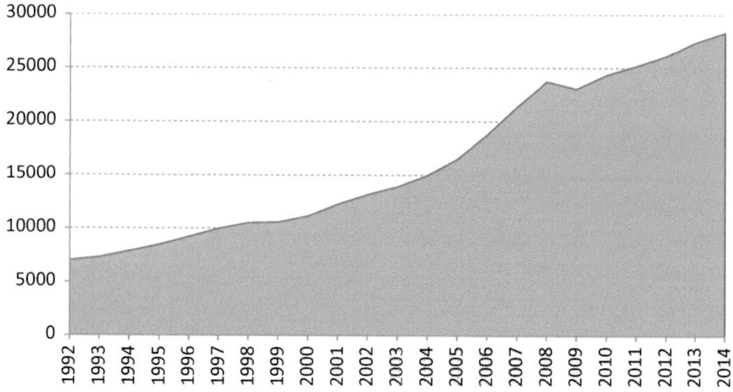

Fig. 3.1 GDP per capita development 1992–2014 in Slovakia (in USD). *Source* OECD 2015 (1992–1994, 2014—estimation)

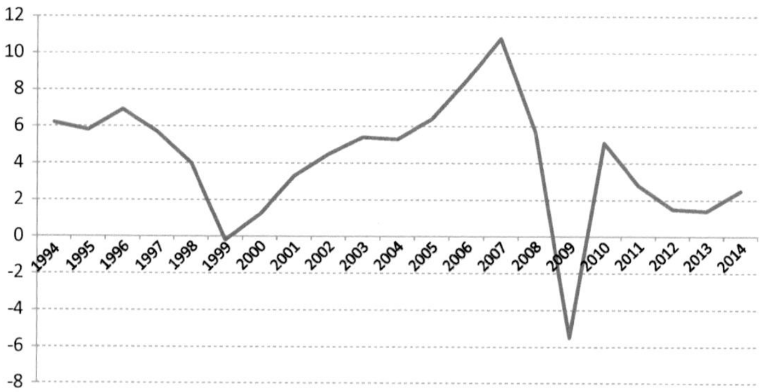

Fig. 3.2 GDP growth in Slovakia 1994–2014 (year to year change in %). *Source* OECD (2016) (1994–1996, 2014 estimated)

Fig. 3.3 Unemployment rate in Slovakia 1991–2014 (in %). *Source* OECD

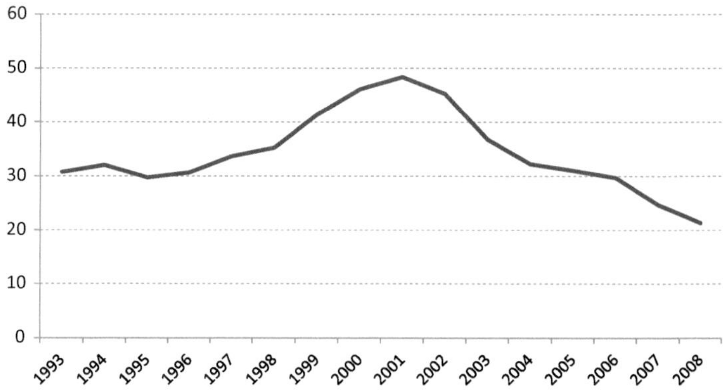

Fig. 3.4 Exchange rate SKK (Slovak Crown) to USD. *Source* NBS 2015

We use a set of longer term basic economic indicators (GDP growth, unemployment and exchange rate development—Figs. 3.2, 3.3, 3.4) to document the economic and social development context of public administration reforms. The brief view indicates many phases of economic growth and economic slowdown, or even economic decline. Early transition public administration reforms had a limited scope and were implemented in a less transformed economy and had a more democratizing nature. The macroeconomic situation worsened in the mid-1990s (e.g. Marcinčin and Beblavý 2000). The economy lost its dynamics, which was visible in the higher unemployment rate (1994–1995) and GDP growth decline. Short-term renewal of economic growth was achieved only thanks to increased state involvement in the economy, with rising external and internal debt (e.g. MESA 1998). Real economic restructuring and reform was absent and economic imbalance expanded. In such a situation, only state administration (de-concentration) and territorial division-based reform were implemented. Bad economic development and many other factors caused the fall of the government in 1998 parliamentary elections.

The Slovak economy situation during the years 1998–2001 was one of the most complicated in its history. The entire economic and social environment had been disrupted. New political garniture recognized that the only way out of this situation was in wide-scale social and economic reforms. Such an inevitable reform drive also had support among citizens and key societal actors. Public administration reform was incorporated into the main package of reforms. The new economic institutional framework also required changes in public administration (e.g. to implement fiscal decentralization). Decentralization, as one reform priority, included the introduction of regional self-government, which had also been considered as a tool for a new approach to development within the country. The previously prevailing dependency on central state activity in regional and local development needed to be replaced by more powers and resources allocated to the sub-state levels of government. It should motivate them for a more initiative approach to development of their territories. After this period, local self-governments (e.g. in large cities) and regions started to be more active actors in the economic and social development of the country.

Positive macroeconomic development after successful reforms, combined with a general positive economic cycle, finished with the start of the financial and economic crisis in 2008–2009. The extremely open Slovak economy could not avoid the consequences of this global crisis, despite limited internal sources of crisis phenomena. The economic crisis that hit a set of important economic sectors was later converted into a public finance crisis (Buček 2012). The effort to cope with the crisis led to excessive budget deficits and increased public debt. Originally negotiated minor adaptation and short-term measures with a limited impact on local and regional self-government (2009) turned into longer term serious interventions into previously functioning local self-governments (e.g. Buček and Sopkuliak 2014).

3.4.5 Public Finance Development

Worse economic development and too expansive government spending can cause a voluntary or forced public finance consolidation policy. Such development puts into question many reform achievements or has changed the nature of reforms under the pressure of necessary measures initiated by public finance development. It has been a very sensitive field for local self-governments' functioning and local development in general. The Slovak case indicates that the main packages of reforms responded also to worse public finance development. The public finance framework had been important already during the early transition years' reforms. These early reforms addressed primarily the need for democratization at the local level, but the unpredictable transitional public finance situation limited any deeper reforms, e.g., leading to more extensive decentralization. The reform that started at the end of 1990s was more complex and progressive (e.g. increased financial capacities at the local level), while reform steps after the economic and financial crisis were more "restrictive" measures (partly interim), with "counter-reform" features (e.g. limited financial autonomy at the local level). Public finance consolidation measures in this case diminished the effect of previous reforms and circumscribed more progress in reforms, although selected modernization steps could be taken.

Public sector savings or an increase in taxes (as well as tax collection improvement) is usually at the core of a consolidation effort. It can result in larger scale public administration reforms, or at least interim consolidation fiscal measures. Primary goals usually include reduction of public debt and at least the freezing of public administration expenditures. If we take the public debt rise, e.g. compared to GDP, we can observe two key periods with a quick and extensive rise in Slovakia after 1990 (Fig. 3.5). The first period culminated in the years 1998–2000. A longer period of public debt decrease from 2000 to 2008 reflected positive economic development and many reforms were introduced, including public administration reform. This was a very progressive reform period also from the public finance point of view (e.g. banking sector consolidation and privatization, tax reform), with an increase in fiscal capacity at the local level (fiscal decentralization, but including limits on local borrowing). This successful consolidation period followed several years with more vigorous public finance, influenced by effort to meet Eurozone criteria.

A new wave of public debt growth emerged in 2008. As OECD (2014) outlined, high budget deficits and growing debt have created the need for fiscal consolidation. Without additional policy interventions, general government debt would continue to increase rapidly. As a result of its public finance development, the Slovak Republic has been included in the Excessive Deficit Procedure under the EU Stability and Growth Pact. The crisis years meant increased instability, less predictability and more subordination of sub-state public finance to national public finance priorities. As Buček and Sopkuliak (2014) revealed—many of the measures induced by the financial and economic crisis and later on induced by fiscal consolidation focused on local and regional self-governments. It included a decreased

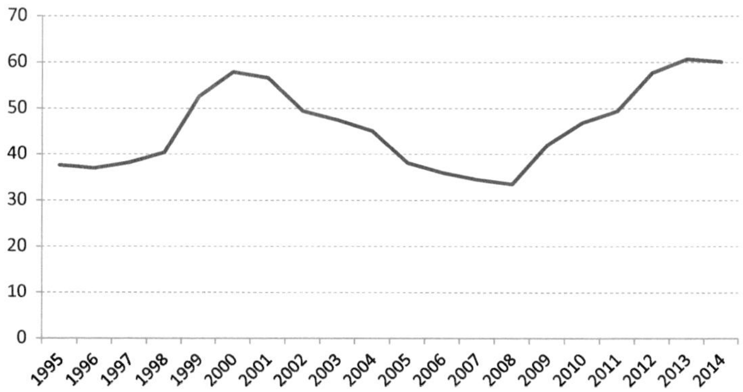

Fig. 3.5 General government debt to GDP development in Slovakia (in %). *Source* OECD (2016)

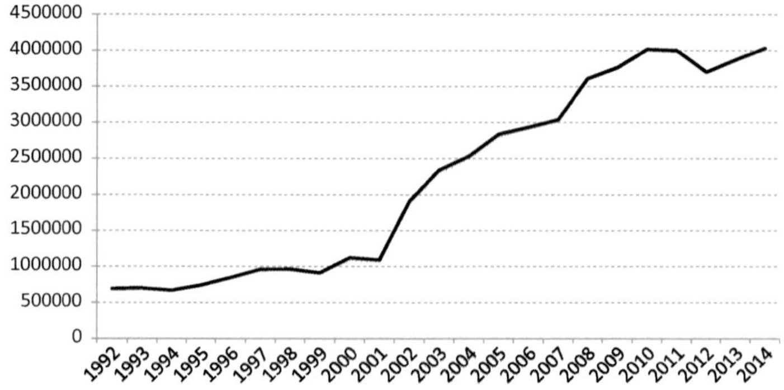

Fig. 3.6 Total revenues of local self-governments in Slovakia (th. EUR). *Sources* Ministry of Finance of the Slovak Republic (2016)

portion of shared taxes (in favour of the central state), voluntary savings compared to previous budgets, pressure to reduce debt financing (borrowing, other external forms of financing). They were incorporated into public finance consolidation measures that were, in fact, of a counter-reform character. The main state subvention flows were stable, e.g. for education. Nevertheless, this consolidation effort was successful (including the withdrawal of Slovakia from the Excessive Deficit Procedure), also thanks to positive macroeconomic development. Many measures of a counter-reform nature were later mitigated, and a pre-crisis framework was re-established for 2015–2016. It included a return to better shared tax distribution formulas strengthening the position of self-governments.

The specific context of reforms at the local level from the public finance point of view is shown in Fig. 3.6. While reforms after the year 2000 had a very positive impact on local finance, "defensive" reforms decided on during post-crisis public finance consolidation had a contradictory effect with a slowdown of development at the local level (2011–2013). The absence of a local prodevelopment base of this reform started to be restored by a return to the pre-crisis framework in 2014–2015.

3.4.6 Central Governments, Leading Political Parties and Main Policy Priorities

The role of central governments in local public administration reforms has been very influential in Slovakia. Short-term duration, instability, internal diversity and competing policy priorities of central governments reduced the chances to prepare more elaborated public administration reforms during the first half of the 1990s. They have not been able to adopt a serious decision, e.g. concerning decentralization, or the meso-level of government (besides limited progress within early transition reform). All central governments worked only for short periods (see

Table 3.2)—during the first 4 years 1990–1994, the governments of Prime Ministers M. Čič (transitional—December 1989–June 1990), V. Mečiar (June 1990–April 1991, Mečiar I), J. Čarnogurský (May 1991–June 1992), again V. Mečiar (June 1992–March 1994, Mečiar II) and J. Moravčík (March 1994–December 1994) were in power. They were also coalition governments with an absence of strong internal cohesion. Many governments announced plans for reforms, established special bodies and initiated various analytical and conceptual works on the future of public administration (e.g. two documents focusing on the new organization of local public administration and concerning new administrative and territorial division in 1993). The instability of central government and their changing approaches and priorities led to a slowing down of reforms, or short-term changes (mainly in reorganization of state administration with signs of de-concentration, see e.g. Slavík 1997).

The next central government of Prime Minister V. Mečiar (1994–1998, Mečiar III.) governed a full 4-year term for the first time in Slovak political history. However, V. Mečiar was a controversial, strong leader with his own visions concerning public administration and its reform (reflecting also the centre-left and nationalist party composition of this government coalition). His government focused on stronger state administration as an inevitable part of its perception of new state-building processes. Although, there were possibilities to discuss reforms proposed during this government, the final decision was primarily political and partly unexpected. It reflected the opinions and interest of the governing coalition. The reform introduced in 1996 was criticized in many fields. It was an especially vague strategy concerning the division of powers and political decisions on territorial administrative division (without a search for a larger consensus). The main obstacle had been the strong emphasis on state administration and de-concentration, without a clear relationship towards decentralization (which had been promised for the future). According to Mesežnikov (2002), this government focused on strengthening its own position, including strengthening the position of its own supporters (their political parties' regional and local elites).

The governments of Prime Minister M. Dzurinda (1998–2002 Dzurinda I. and 2002–2006 Dzurinda II.) represented a democratic coalition that defeated V. Mečiar. It had a strong mandate for reforms. The central government was formed by a group of political parties representing centre-right (SDK and SMK) as well as centre-left (SOP) and left parties (SDL) during the first term, while the coalition during the second term had a more centre-right profile. These two governments we can consider as the most reform-oriented in Slovak history until 2015. They prepared and realized more complex reforms, including the most important public administration reform at the sub-state level. As far as public administration reform was concerned, it obtained political priority, including the introduction of the post of governmental plenipotentiary for public administration reform and decentralization. This government considered the previous reforms of state administration as being incomplete (1995–1996), too costly, in many features outdated, and with need of serious modification. A strong driving force was the process of EU integration (Slovakia joined the EU in 2004) and related conditionality. However, due

Table 3.2 Central governments and their composition in Slovakia 1989–2016

Prime minister	Duration	Political parties	Profile
M. Čič	1989–1990	VPN—Public Against Violence Movement	Transitional
V. Mečiar (I.)	1990–1991	VPN—Public Against Violence Movement, KDH—Christian Democratic Movement, DS—Democratic Party	Centre-right
J. Čarnogurský	1991–1992	VPN—Public Against Violence Movement, KDH—Christian Democratic Movement, DS—Democratic Party, MNI—Hungarian Independent Initiative	Centre-right
`	1992–1994	HZDS—Movement for Democratic Slovakia, SNS—Slovak National Party	Centre-left
J. Moravčík	1994	DÚ—Democratic Union, KDH—Christian Democratic Movement, SDĽ—Party of Democratic Left	Centre-right
V. Mečiar (III.)	1994–1998	HZDS—Movement for Democratic Slovakia, ZRS—Association of Workers, SNS—Slovak National Party, RSS—Slovak Farmers Party	Centre-left
M. Dzurinda (I.)	1998–2002	SDK—Slovak Democratic Coalition, SDĹ—Party of Democratic Left, SMK—Party of Hungarian Coalition, SOP—Party of Civic Understanding	Centre-right
M. Dzurinda (II.)	2002–2006	SDKÚ—Slovak Democratic Christian Union, ANO—Alliance of New Citizen, SMK—Party of Hungarian Coalition, KDH—Christian Democratic Movement	Centre-right
R. Fico (I.)	2006–2010	SMER—Social Democracy, SNS—Slovak National Party, ĽS-HZDS—People's Party-Movement for Democratic Slovakia	Centre-left
I. Radičová	2010–2012	SDKÚ—Slovak Democratic Christian Union, SaS—Freedom and Solidarity, KDH—Christian Democratic Movement, MOST-HÍD—Bridge—Civic Party	Centre-right
R. Fico (II.)	2012–2016	SMER—Social Democracy	Centre-left

Slovak usual abbreviations are combined with full English names of parties

to the inconsistency of the governmental coalition, not all intentions were achieved as planned (e.g. in the field of territorial administrative division). Centre-left and left parties limited further progress in some parts of the reform (mainly during the end of the 1998–2002 term). They were motivated by fears concerning the weakening of their position in local public administration. Thanks to the re-election of the leading political parties of this coalition (2002), major changes were implemented, including fiscal decentralization applied after 2005. Focus on implementation of this wide-scale reform limited preparation of further stages of local public administration reforms, including growing tensions within the governing coalition in 2005–2006 (with the political parties Alliance of New Citizen and the Christian

Democratic Movement). Despite preparation of a reform document focusing on reorganization of local self-government for public discussion (Office of the Government of the Slovak Republic 2004), it lost priority and went into the background of this government's agenda.

The primarily centre-left (SMER—Social Democracy) government of Prime Minister R. Fico (2006–2010) did not continue previous reforms. The main effort of this government was fulfilment of the criteria for joining the Eurozone (Slovakia joined the Eurozone in 2009). During the prevailing positive phase of the economic cycle, it could also realize a more typical social democratic policy focusing on unemployment and the better social situation of citizens. Later on it dealt with the impact of the financial and economic crisis in 2009–2010. Measures adopted in 2009 worsened the financial situation of local self-governments and many of them had to adopt their own preventive measures (Buček and Sopkuliak 2014). This government did not show any interest in any kind of deeper reform. Another major change was the further reorganization of state administration, including elimination of regional offices of general state administration. In the field of local self-government, this internally diverse government concentrated on maintaining the existing framework with minor changes incorporated under the "headline" of modernization (mainly ICT application expansion). During the second part of its term, the central government initiated preparation of a new concept of reform of self-government, which had been submitted in its final form in July 2009 (Office of the Government of the Slovak Republic 2009). It included the intention to focus on more efficient public administration (including joint offices for more self-governments), human resources improvement, information technologies application and better monitoring of local self-government. More costly tasks would be financed by the resources of the EU Operational Programmes 2007–2013. The end of the electoral term stopped further progress and this government was not re-elected.

The centre-right government of PM I. Radičová (2010–2012) was deeply involved in formulating anti-crisis measures also concerning local self-government, e.g., cuts in spending, decrease of their share of PIT yield, limits on their debt. Nevertheless, this government declared its support for further reforms in public administration in its government programme manifesto. It included certain features usually mentioned within repeated attempts to prepare "communal reform" (e.g. more motivating conditions for voluntary inter-municipal co-operation, joint administration of selected powers, or even amalgamation of small local self-governments, Office of the Government of the Slovak Republic 2010). This government started again with preparation of a conceptual document for reform. Despite the fragments of this concept, it was not completed due to the early fall of this government. Nevertheless, this government expressed interest in deeper reforms, with many reform proponents from years 1998–2006 in important political positions at the central level.

The second government led by PM R. Fico (2012–2016) as a "single colour" government had a great opportunity for more extensive reforms. Nevertheless, it focused on reforms of state administration known as the ESO Programme, focusing on improving the delivery of administrative services, with their strengthened quality, better access (as "one-stop shops"), improved cost-efficiency and savings and integration of dispersed specialized state administration offices. The large-scale audit of public administration finished without a clear policy outcome. This government declined any deeper reform of self-government. Its effort in this field focused on modernization changes. Despite a declaration (Office of the Government of the Slovak Government 2012), no substantial progress had been made in the field of "joint offices" serving more self-governments. This government was more successful in its support of e-government activities within self-government. This government concentrated more on fiscal consolidation issues at all levels and lines of public administration and a more typical social policy. It is well documented by the speech of the PM R. Fico at the 25th ZMOS Congress in June 2014: "*Ladies and gentlemen, as I already said, during 2012, 2013, and unfortunately also during the first half of 2014, in our society only one word dominated—money. Finance, sustainable finance, public order, consolidation, these are words and slogans, which are the most frequent ... Words, which are painful for the state sector, as well as self-governments*" (Obecné Noviny 2014, p. 8).

When considering public administration reform, we cannot underestimate the influence of policy priorities of central governments (see Fig. 3.7). Key critical junctures were 1989 as the fall of the Communist regime, 1993 as the origin of the Slovak Republic as an independent state, or the year 2004 when Slovakia joined the EU and related processes. Policy priorities strongly influenced the willingness and capacity to prepare and implement public administration reforms. Administrative reforms were not always necessary to implement political priorities successfully in a particular period. From the point of view of self-government, the most important reform periods according to political priorities were the introductory democratization period and the period prior to joining the EU. Other periods were less reform progressive, or focused more on state administration reforms.

Fig. 3.7 Policy priorities in Slovakia after 1989 (own elaboration)

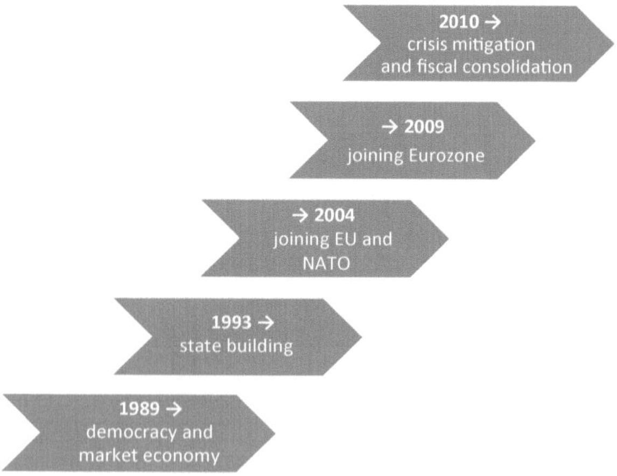

The concentration of central government capacities on political priorities and related changes has limited the ability to move ahead with local public administration reforms. It is natural especially if implementation of particular policies is strongly linked to particular sections of the public administration system (that should be stable to remain efficient enough to implement changes). Many reforms do not always require deeper changes in local public administration. For example, there was little attention to public administration reform after joining the EU and later during the effort to join the Eurozone, combined with an orientation towards administrative capacities, e.g., on EU funds absorption, or administration processes innovation such as programme budgeting in local self-governments. We can also observe periods with less reforms, focusing on adaptation to previous changes and eventual minor improvements that are sometimes needed (after experience with reform outcomes).

The impact of political parties has been crucial on the direction, scope and timing of reforms in Slovakia. As various political parties have had access to central political power, they also differently influenced the dynamics of reforms. Parties had different opinions on the level of centralization or decentralization, or the role of state administration and self-government within public administration, territorial administrative division and so on. Besides the composition of the central government, especially during later stages of the post-socialist transition, another important interrelated aspect was the duration of central governments. Short-term governments and single-term governments were less able to prepare and implement reforms, especially those concerning self-government. Such reforms require more than one electoral period. The implementation of reforms was concentrated on in the second periods (Dzurinda II, Fico II—in this case after the short-term break 2010–2012, and to a certain extent also Mečiar III).

There were established political parties, usually active for a long time in political life, repeatedly participating in central governments that put into practice more extensive local public administration reforms. This reflects the need for existence, capacities and leadership of large and influential political parties, although in practice operating often in a coalition with smaller parties. Such a scale of reform is not an easy task and more capacities are needed, including necessary respected professional capacities available in well-established parties with experience in participating in central government and in managing changes. It also needs the capacity and willingness to build a larger consensus across many segments of society.

It is less simple to summarize the ideological orientation of political parties and their reform inclination. Experience from the last 25 years indicates more willingness to prepare and implement reforms by centre-right governments in Slovakia. Centre-left governments paid less attention to these issues. It also seems that centre-left and left political parties (Mečiar III, Fico I and Fico II) focused more on reform of state administration, and a less decentralized and stronger state. It is easier to implement, with less preparatory works, and less extensive coordination with social partners and public consultations. Nevertheless, in this we have to be careful, having less time for evaluation and drawing far-reaching conclusions.

3.4.7 Governance and Other Actors

Slovak society and its development are no longer dominated exclusively by central government. The central state has been consulting its intentions with other actors active in social, economic and political life to a growing extent. It means that final decisions concerning any reforms are debated not only in central state institutions and within governing political parties. A certain level of consensus or compromise in reforms had been achieved after 1998. In the case of local public administration reforms, representatives of local self-government associations held a key role, but other actors also have much to say (trade unions, representatives of employers, third sector bodies). The governance structures include official bodies like the Economic and Social Council of the Government, or special governance-based bodies affiliated to particular ministries. Until 2011, among the key institutions related to reforms was the Council of the Government for Public Administration (as an advisory and consulting body), with representatives also outside state administration.

Since the early period of post-socialist transition, we can observe the influential and growing role of representatives of local self-governments. A key partner of central governments is primarily the Association of Towns and Communes in Slovakia (in Slovak ZMOS—Združenie miest a obcí Slovenska), representing about 95 % of all local self-governments (one of the most influential policy actors outside of central government and political parties in Slovakia). This association actively negotiates in favour of its members (e.g. including very small communes). Similar associations—Union of Cities (in Slovak—Únia miest), as well as the Association of Self-governing Regions (in Slovak—Združenie samosprávnych krajov), are less influential, although respected. Nevertheless, while ZMOS, in the early stages of transition had been a strongly proreform-oriented actor, later attitudes have documented a less enthusiastic approach. Its considerations fluctuate mostly within the local self-government framework achieved during reforms implemented during the first half of the decade after 2000. It focuses on protection of already obtained positions and seems less open to new reforms, especially more radical ones—like amalgamation (which could threaten the position of many mayors), or obligatory or forced co-operation in the case of smaller local self-governments. However, this large association with large internal diversity is not easy to move forward into general acceptance for more advanced reforms. It is also less active in generating its own larger reform activities, working mostly on improvements to the existing framework for local self-government. On the other hand, in many cases, representatives of local self-governments were able to stop or moderate reforms they strongly opposed, e.g. transfer of particular powers, changes in the tax yield distribution system (e.g. Buček and Sopkuliak 2014).

During the last 25 years, many other institutions participated in the reform effort in various ways. For example, many experts from Slovak universities participated in analytical and conceptual works (e.g. Slavík et al. 2005). Important contributions were made by the non-governmental sector. While immediately after the change of regime this sector only just started to grow, the vital non-profit non-governmental sector already existed during the second half of the 1990s (partly as opposition to the then central government of PM V. Mečiar). Among the NGOs that had a long-term interest in public administration reform we have to mention the MESA 10 think tank (e.g. Mikloš et al. 1998), which was influential especially by its involvement in decentralization reform preparation. Representatives of this NGO later directly participated in reform implementation after 1998. There are also other current NGOs still advocating public administration reforms primarily based on inter-municipal co-operation and amalgamation (following the ideas of municipal reform), or evaluating the effects of previous reforms. This is the case with the M.R. Štefánik Conservative Institute (Slov. Konzervatívny inštitút M.R. Štefánika) and the Municipal Research and Advisory Centre. Together with academic institutions, they provide many interesting source materials (e.g. Sloboda 2010; Komunálne výskumné a poradenské centrum 2014) contributing to the shaping of potential future reforms.

3.4.8 Coincidence of Events and Factors Influencing Local Public Administration Reforms

Coincidence of main reforms periods, key social and political events, factors development and their linkage provides good opportunity to get closer to answer main research question—Why and when local public administration reforms are prepared and implemented? Brief overview of selected factors provides coincidence matrix (Table 3.3). As "critical junctures" influencing reforms dynamic and scope, we can consider fall of Communist regime in 1989 and joining EU in 2004. They induced the most important reforms (after 1989, and prior to 2004). As secondary events with minor influence, we can also add Slovak Republic origin in 1993 and joining Eurozone with weakening effect on reform effort in local self-government (after 1993, prior to 2009).

Due to the dual model of public administration applied in Slovakia, we can observe reform dynamics that are different between local self-government and local state administration. Reforms are implemented in parallel, as well as in a separate way. While reforms in state administration are more permanent and incremental, reforms in self-government are implemented in more identifiable stages (e.g. after 1989, at the turn of the century, or induced by the financial and economic crisis). A reform of state administration is easier to prepare and implement within central government's own decision-making framework. On the other hand, reforms of self-government need more governance-based decision-making with a long-term search for consensus on changes in wider scale.

If we want to evaluate the role of factors that influence local public administration reforms, we have to mention their different strength. Some of these appear to be very significant and substantially influenced motivation for local public

Table 3.3 Coincidence matrix of selected events and factors development

	1989	1990	1991	1992	1993	1994	1995	1996	1997	1998	1999	2000	2001	2002	2003	2004	2005	2006	2007	2008	2009	2010	2011	2012	2013	2014
Reform of																										
Self-government		o	o								o	o	o	o	o	o	o				o	o	o	o	o	
State administration		o	o	o	o		o	o			o	o	o	o	o	o			o					o	o	o
Key events																										
Year of event*	o				o											o					o					
Central government																										
Centre-left				o	o	o	o	o	o	o								o	o	o	o	o		o	o	o
Centre-right		o	o	o		o				o	o	o	o	o	o	o	o	o				o	o	o		
Government second/third period						o	o	o	o	o				o	o	o	o	o						o	o	o
Macroeconomic and public finance indicators																										
GDP growth**											o	o									o			o	o	
Unemployment rate***						o	o				o	o	o	o	o	o	o	o			o	o	o	o	o	o
Public debt to GDP****											o	o	o	o	o	o						o	o	o	o	o

* Key events: 1989—fall of communist regime, 1993—origin of the Slovak Republic, 2004—joining EU, 2009—joining Eurozone, start of economic and financial crisis in Slovakia (e.g. for 1989 only key event is mentioned)
** GDP growth—below 2 % year to year, according to OECD 2016
*** Unemployment rate—above 13 % according to Statistical Office of the Slovak Republic 2015
**** Public debt—above 45 % of GDP according to OECD 2016

administration reform (besides internal needs for reform, e.g., as modernization). Macroeconomic development and public finance are core factors that influence the motivation for reform. Nevertheless, it seems that under economic and fiscal pressure we can expect administrative reforms of various kinds—more sophisticated deeper reforms with a long-term positive impact, and less sophisticated (disappearing after some time)—narrowly focusing, for example, on expenditures savings in the short-term view. To the core factors, we can also add interrelated factors combining the role of central governments and political parties. It seems that political stability supported the chance of successful reforms. Reforms have a better chance if there are stable (more electoral terms in government) and well-established political elites. Besides early transition public administration reform, subsequent reforms were performed by the governing coalition with a clear leading party (SDKÚ), and/or by a single party government (SMER–SD). They had the capacity to prepare and implement reforms, having had longer term experience in central government participation and leadership. Reform capacities and vision seem better developed in the case of centre-right parties (following a liberal agenda), and less developed in the case of centre-left parties (putting into question the situation concerning the development of leftist thinking in this part of Europe). They have different political priorities and they have paid different attention to local self-government comparing to local state administration.

On the other hand, some of the discussed factors were influential only in a particular period and later on their impact decreased, or their nature changed. This is, for example, the case with administrative traditions and policy legacies (but they can always return into discussion). The modernization "argument" for reform has also changed—now strongly moved in favour of its "technological" and managerial meaning in Slovakia. International influences also lost their strength since the time of pre-accession processes, although many piecemeal policy transfers are permanent. Surprisingly less reform calls have been generated from within local public administration, including more autonomous local self-government elite. This elite lack ability in generating larger scale proposals, although they are quite efficient in protecting previous reform achievements and in proposing minor improvements to the public administration system. Compared to large associations, other NGOs are less influential, although they provide important knowledge and alternative solutions.

3.5 Conclusions

Reforms in general are an immanent part of any societal progress. Public administration reforms are among the most typical expressions of such development from the point of view of governments and the public sector. Although there are constant debates on the weaknesses and strengths within local public administration, and on needed improvements and modernization, it is often not enough for larger scale reform implementation. A previous overview of development in Slovakia from a longer term perspective shows that reforms usually need more complex stimuli. A combination of need for progress in the field of local public administration and local development (democratization, decentralization, modernization), less positive indicators of social and economic development (such as during post-Communist transformation, economic and financial crisis), as well as the capacities and selected characteristics of central governments and leading political parties, plays a primary role.

If we attempt to evaluate the long-term trend in reforms, we can see respect paid to the continuity, progress and moderate modifications of previous reforms in Slovakia. Despite the contradictory positions of political parties during the various local public administration reforms, there were not any serious interventions into previous self-government reforms after changing political parties in government. Although modifications emerged, no dramatic irreversible counter-trend has been observable. It reflects a certain kind of consensual approach, partly caused by the coalition nature of most central governments and a more extensive consultation on such changes with other societal actors in Slovakia (with the respected position of representatives of local and regional self-government). Even in the case of intervention into certain local self-governments' rights and resources allocation adopted during the period of fiscal consolidation, promises that after stabilization of public finance the "rules of the game" would be returned to the previous state are fulfilled (e.g. since 2016 in selected measures). Nevertheless, changes in the field of local state administration are less unidirectional.

Preparation and implementation of any kind of future reforms within Slovak local public administration (and especially in local self-government) are more complicated issues. The current mode of governance in managing important multilevel political issues limit quick progress in building a consensus around any new round of reforms. Nevertheless, it is also clear that at least one deeper local self-government reform is still needed. Local self-government reform is behind the changes in state administration, it is fragmented, and many tasks cannot be provided efficiently by small local self-government (e.g. ICT application in local self-government units below 200 inhabitants). Integration or a joint service provision base seems inevitable. It looks like the applied incremental changes to the present are not able to solve a serious part of local self-government problems. Among the more serious obstacles, we can consider only the slowly changing opinions of local self-governments expressed by their leading association (ZMOS). It is based on internal solidarity with the interests of small local self-governments in mind. It limits progress in the most serious issue concerning the future of small local self-governments. This association, which has played a progressive role in many other issues, is moving only very slowly in finding a more efficient and flexible solution to the problem of small local self-governments. Nevertheless, it is a very influential actor and any progress in reforming local self-government will not be possible without changing their attitudes. As a result, as a first midterm scenario, we can expect minor gradual reforms focusing on improvements in the functioning of local self-government, including a potential further slow transfer of powers and related resources in favour of local self-governments, which will have impacts on urban governance processes. The less probable second midterm scenario is that long-term public pressure reflecting the inefficiencies on the functioning of existing local self-government, combined with a rising political will for reform, can lead to more extensive reforms in the field of self-government focusing on local as well as meso-level of self-government.

Acknowledgments The chapter was prepared with thanks to support provided by the project APVV-018-12 entitled "Human Geographical and Demographic Interactions, Nodes and Contradictions within a Time-Space Network" (50 %) and VEGA research Grant No. 1/0745/16 "Autonomy, Interdependence and Interactions of Spatial Systems" (50 %).

References

Act No. 369/1990 Coll. on local government (as amended)
Act No. 517/1990 Coll. on territorial and administrative division
Act No. 138/1991 Coll. on municipal assets (as amended)
Act No. 221/1996 Coll. on territorial and administrative organisation
Act No. 222/1996 Coll. on organisation of local state administration (as amended)
Act No. 302/2001 Coll. on self-government of higher territorial units
Act No. 416/2001 Coll. on transfer of powers from state administration to communities and higher territorial units
Act No. 515/2003 Coll. on regional offices and district offices
Act No. 595/2003 on income tax as amended on 1 Dec 2011
Act No. 583/2004 Coll on budgetary rules of territorial self-government (with its amendments)
Act No. 254/2007 Coll. on regional offfices cancellation
Act No. 345/2012 Coll. on selected measures in local state administration
Act No 180/2013 Coll. on organization of local state administration (as amended)
Alba CR, Navarro C (2011) Administrative tradition and reforms in Spain: adaptation versus innovation. Public Adm 89(3):783–800
Aucoin P (1990) Administrative reform in public management: paradigms, principles, paradoxes and pendulums. Governance 3(2):115–137
Bakota B (2014) Local and regional government reform in Croatia: subsidiarity and innovation in an era of austerity. In: Silva CN, Buček J (eds) Fiscal austerity and innovation in local governance in Europe. Ashgate, Farnham, pp 113–128
Baldersheim H (2014) Reflections on 25 years of local government reforms in Europe. In: Monsen LK, Otterlei JB (eds) Valg, velferd og lokaldemokrati: Festskrift til Per Arnt Pettersen. Akademika, Bergen, pp 17–34
Blöchliger H, Song D, Sutherland D (2012) Fiscal consolidation: part 4. Case studies of large fiscal consolidation episodes, OECD Economics Department Working Papers, 935, OECD Publishing, Paris
Bothe W (1955) Coincidence method. Science 122:861–863
Bouckaert G, Nakrošis V, Nemec J (2011) Public administration and management reforms in CEE: main trajectories and results. NISPAcee J Public Adm Policy 4(1):9–29
Brown D (2005) Electronic government and public administration. Int Rev Adm Sci 71 (2):241–254
Bryson PJ (2008) "State administration" vs. self-government in the Slovak and Czech Republics. Communist Post Communist Stud 41(3):339–358
Buček J (1993) Problémes de financement municipal en Slovaquie. Bull Soc Languedoc Geogr 116(3–4):335–345

Buček J (2006) The position of local self-government after main phase of public administration reform in Slovakia. Acta Geogr Univ Comen 48:167–178
Buček J (2011) Building of regional self-government in Slovakia: the first decade. Geografický časopis 63(1):3–27
Buček J (2012) Crisis in Slovakia 2009–2010: from saving the economy to saving public finance. In: Gorzelak G, Goh CC, Fazekas K (eds) Adaptability and change: the regional dimensions in central and eastern Europe. Wydawnictwo Naukowe Scholar, Warszawa, pp 334–359
Buček J, Sopkuliak A (2014) Local self-government finance in Slovakia during the 2008–2011 crisis. In: Silva CN, Buček J (eds) Fiscal austerity and innovation in local governance in Europe. Ashgate, Farnham, pp 7–30
Búšik J (2005) Reformné procesy v štátnej správe. In: Žárska E, Šebová M (eds) Decentralizácia verejnej správy Slovenskej republiky: otvorené otázky. Ekonomická Univerzita, Bratislava, pp 8–18
Capano G (2003) Administrative traditions and policy change: when policy paradigms matter. The case of Italian administrative reform during the 1990s. Public Adm 81(4):781–801
Capoccia G, Kelemen RD (2007) The study of critical junctures: theory, narrative, and counterfactuals in historical institutionalism. World Polit 59(3):341–369
Connaughton B (2008) Ireland: modernization as opposed to radical reform. In: Killian J, Eklund N (eds) Handbook of administrative reform: an international perspective. Auerbach Publications, Boca Raton, pp 95–113
Conrad CR, Golder SN (2010) Measuring government duration and stability in Central Eastern European democracies. Eur J Polit Res 49(1):119–150
Dellepiane-Avellaneda S, Hardiman N (2015) Fiscal politics in time: pathways to fiscal consolidation in Ireland, Greece, Britain, and Spain, 1980–2012. Eur Polit Sci Rev 7 (2):189–219
Demmou L, Price RW (2015) Reforming the Slovak Public Sector. OECD Economics Department Working Papers No. 1212, OECD Publishing, Paris. doi:10.1787/5js1pzrvxmkg-en. Accessed 17 Sep 2015
Dolowitz DP, Marsh D (2000) Learning from abroad: the role of policy transfer in contemporary policy-making. Governance 13(1):5–23
Dunleavy P, Margetts H, Bastow S, Tinkler J (2006) New public management is dead—long live digital-era governance. J Public Adm Res Theory 16(3):467–494
Fidrmuc J (2003) Economic reform, democracy and growth during post-communist transition. Eur J Polit Econ 19(3):583–604
Grabbe H (2001) How does Europeanization affect CEE governance? Conditionality, diffusion and diversity. J Eur Public Policy 8(6):1013–1031
Hall PA (2010) Historical institutionalism in rationalist and sociological perspective. In: Mahoney J, Thelen K (eds) Explaining institutional change: ambiguity, agency, and power. Cambridge University Press, Cambridge, pp 204–224
Hibbs DA (1977) Political parties and macroeconomic policy. Am Polit Sci Rev 71(4):1467–1487
Illner M (2003a) Devolution of government in the ex-communist countries: some explanatory frameworks. In: Baldersheim H, Illner M, Wollmann H (eds) Local democracy in post-communist Europe. Leske+Budrich, Opladen, pp 9–28
Illner M (2003b) The Czech Republic 1990–2001. Successful reform at the municipal level and a difficult birth of the intermediary government. In: Baldersheim H, Illner M, Wollmann H (eds) Local democracy in post-communist Europe. Leske+Budrich, Opladen, pp 61–90
Kaczmarek T (2005) Struktury terytorialno-administracyjne i ich reformy w krajach europejskich. UAM, Poznan
Klimovský D (2008) Politics and its impact on the reform processes: the case of public administration reform in Slovakia (1989–2006). In: Musil J (ed) Space and historical time as dimensions of social change. Matfyzpress, Prague, pp 45–64
Klimovský D (2014) Intermunicipal cooperation in Slovakia: the case of regions with highly fragmented municipal structure. Faculty of Organization Studies, Novo mesto

Komunálne výskumné a poradenské centrum (2014) Zlučovanie a spolupráca obcí. Piešťany. http://www.komunal.eu/sk/. Accessed 17 Jan 2016

Kovács IP (2009) Europeanisation of territorial governance in three Eastern/Central European countries. Adm Cult 10:40–57

Ladi S (2014) Austerity politics and administrative reform: the Eurozone crisis and its impact upon Greek public administration. Comp Eur Polit 12(2):184–208

Magone JM (2011) The difficult transformation of state and public administration in Portugal. Europeization and the persistence of neo-patrimonialism. Public Adm 89(3):756–782

Mahoney J, Thelen KA (2010) Theory of gradual institutional change. In: Mahoney J, Thelen K (eds) Explaining institutional change: ambiguity, agency, and power. Cambridge University Press, Cambridge, pp 1–37

Malíková Ľ, Jacko T et al (2013) New public management a Slovensko: Podoby a aplikácia princípov. Univerzita Komenského, Bratislava

Marcinčin A, Beblavý M (2000) Hospodárska politika na Slovensku 1990–1999. Centrum pre spoločenskú a mediálnu analýzu, Bratislava

Mathernová K, Renčko J (2006) Reformology: the case of Slovakia. Orbis, Fall, pp 629–640

MESA 10 (1998) Sľuby a realita: Slovenská ekonomika 1995–1998. MESA 10, Bratislava

Mesežnikov G (2002) Činnosť politických strán a reforma verejnej správy. In: Mesežnikov G, Nižňanský V (eds) Reforma verejnej správy na Slovensku 1998–2002. Inštitút pre verejné otázky, Bratislava, pp 121–146

Meyer-Sahling J (2009) Varieties of legacies: a critical review of legacy explanations of public administration reform in East Central Europe. Int Rev Adm Sci 75(3):509–528

Mikloš I, Nižňanský V, Žárska E (1998) Nový systém financovania verejnej správy. M.E.S.A 10, Bratislava

Ministry of Finance of the Slovak Republic (2008) Národná koncepcia informatizácie verejnej správy. http://www.informatizacia.sk. Accessed 27 Jan 2016

Ministry of Finance of the Slovak Republic (2015) Národná koncepcia informatizácie verejnej správy: novelizácia (November 2015). http://www.informatizacia.sk. Accessed 27 Jan 2016

Ministry of Finance of the Slovak Republic (2016) Final Accounts of the Slovak Republic 1992–2014. Accessed 5 Nov 2015

Molnar M (2012) Fiscal consolidation: Part 5. What factors determine the success of consolidation efforts? OECD Economics Department Working Papers, No. 936, OECD Publishing

Müller-Rommel F, Fettelschoss K, Harfst P (2004) Party government in Central Eastern European democracies: a data collection (1990–2003). Eur J Polit Res 43(6):869–894

Myant M, Drahokoupil J (2013) Tranzitivní ekonomiky: politická ekonomie Ruska, východní Evropy a střední Asie. Academia, Praha

Nižňanský V (2002) Spoločenský, historický a európsky kontext reformy verejnej správy na Slovensku. In: Mesežnikov G, Nižňanský V (eds) Reforma verejnej správy na Slovensku 1998–2002. Inštitút pre verejné otázky, Bratislava, pp 15–57

National Bank of Slovakia (2015) Exchange rates database 1993–2008. http://www.nbs.sk. Accessed 30 Jan 2016

Obecné Noviny (2014) Slovensko svojimi výsledkami potvrdzuje, že konsolidácia nebola samoúčelná. Obecné Noviny 24:8–9

OECD (2014) Slovak Republic: developing a sustainable strategic framework for public administration reform, OECD Public Governance Reviews. OECD Publishing, Paris. doi:10.1787/9789264212640-en

OECD (2016) GDP per capita, GDP year to year change, General government debt to GDP (indicators). doi:10.1787/dc2f7aec-en, doi:10.1787/b86d1fc8-en, doi:10.1787/a0528cc2-en

Office of the Government of the Slovak Republic (2000) Koncepcia decentralizácie a modernizácie verejnej správy. http://www.rokovania.sk. Accessed 30 Jan 2016

Office of the Government of the Slovak Republic (2009) Návrh Koncepcie modernizácie územnej samosprávy v Slovenskej republike. http://www.rokovania.sk. Accessed 30 Jan 2016

Office of the Government of the Slovak Republic (2010) Programové vyhlásenie vlády Slovenskej republiky na obdobie rokov 2010–2014. Accessed 27 Jan 2016

Office of the Government of the Slovak Republic (2012) Programové vyhlásenie vlády Slovenskej republiky na obdobie rokov 2012–2016. http://www.rokovania.sk. Accessed 27 Jan 2016
Office of the Government of the Slovak Republic (2013) Strategic framework of public administration reform in the Slovak Republic. http://www.rokovania.sk. Accessed 27 Jan 2016
Office of the Government of the Slovak Republic (2015) Návrh Štatútu Rady vlády Slovenskej republiky pre digitalizáciu verejnej správy http://www.rokovania.sk. Accessed 30 Jan 2016
O'Dwyer C (2006) Reforming regional governance in East Central Europe: Europeanization or domestic politics as usual? East Eur Polit Soc 20(2):219–253
O'Dwyer C, Kovalčík B (2007) And the last shall be first: party system institutionalization and second-generation economic reform in postcommunist Europe. Stud Comp Int Dev 41(4):3–26
Perotti R (1998) The political economy of fiscal consolidations. Scand J Econ 100(1):367–394
Péteri G, Zentai V (2002) Lessons on successful reform management. In: Péteri G (ed) Mastering decentralization and public administration reform in CEE. Local Government and Public Service Reform Initiative, Budapest, pp 13–30
Peters BG, Pierre J (1998) Governance without government? Rethinking public administration. J Public Adm Res Theory 8(2):223–243
Pierson P (2000) Increasing returns, path dependence, and the study of politics. Am Polit Sci Rev 94(2):251–267
Potrafke N (2012) Political cycles and economic performance in OECD countries: empirical evidence from 1951–2006. Public Choice 150:155–179
PRAVDA (2014) Kaliňák: Reforma ESO priniesla vlani úsporu 130 miliónov eur. 14.01.2014. www.pravda.sk. Accessed 23 Oct 2015
Rodden J, Wibbels E (2010) Fiscal decentralization and the business cycle: an empirical study of seven federations. Econ Polit 22(1):37–67
Silva CN (2010) The E-planning paradigm-theory, methods and tools: an overview. In: Silva CN (ed) Handbook of research on e-planning: ICTS for urban development and monitoring. IGI Global, Hershey, pp 1–14
Silva CN (2014) The economic adjustment program impact on local government reform in Portugal. In: Silva CN, Buček J (eds) Fiscal austerity and innovation in local governance in Europe. Ashgate, Farnham, pp 31–47
Slavík V (1997) Priestorová organizácia verejnej správy v Slovenskej republike 1990–1996. In: Delmartino F, Vermessen E, Miháliková S, Falťan Ľ (eds) Nové podoby verejnej správy: Slovenská a flámska skúsenosť. Sociologický ústav SAV, Bratislava, pp 91–110
Slavík V, Bačík V, Kožuch M, Ragačová M (2005) Analýza mikroregiónov Slovenskej republiky. Prírodovedecká fakulta UK, Bratislava
Slavík V, Klobučník M, Šuvada M (2013) Geografické aspekty reformy verejnej správy v Slovenskej republike s dôrazom na súčasnú etapu. VSEMVS, Bratislava, pp 39–53
Sloboda D (2010) Čakáreň na municipalizáciu. Konzervatívne Listy 4(April):3
Statistical Office of the Slovak Republic (2015) Electoral Statistics. www.statistics.sk. Accessed on 22 Jan 2015
Statistical Office of the Slovak Republic (2015) Unemployment rate statistics. www.statistics.sk. Accessed on 30 Jan 2015
Steinmo S (2008) Historical institutionalism. In: Della Porta D, Keating M (eds) Approaches and methodologies in the social sciences: a pluralist perspective. Cambridge University Press, Cambridge, pp 118–138
Swianiewicz P (2010) If territorial fragmentation is a problem, is amalgamation a solution? An East European perspective. Local Gov Stud 36(2):183–203
Verheijen AJG (2007) Public administration in post-communist states. In: Peters BG, Pierre J (eds) Handbook of public administration. Sage, London, pp 311–319
Wollmann H (2000) Local government modernization in Germany: between incrementalism and reform waves. Public Adm 78(4):915–936
Wollmann H (2012) Local government reforms in (seven) European countries: between convergent and divergent, conflicting and complementary developments. Local Gov Stud 38 (1):41–70

Chapter 4
Local Self-Government in Hungary: The Impact of Crisis

Adrián Fábián

Abstract The economic crisis affected local self-government much more severely in Hungary than in the other OECD Member countries. The basic characteristics of the Hungarian local self-government system are that it provides local public services and is the basic institutional framework for urban governance. State support is the most important financial source of local public services. Before 2008, the amount of state support, together with other municipal revenues, was not sufficient for financing all local government tasks. The municipalities complemented the missing amount from external sources. The financial problems of the Hungarian local self-governments were not caused by the economic crisis; it only increased these problems: the economic crisis had a considerable negative effect on the financing possibilities of the Hungarian local self-governments. The dramatic increase in the volume of local self-government debt started in 2006 and was primarily caused by the issue of local self-government bonds. This study deals with the related legal background, the economic effects, the reforms for solution and the consequences. The purpose of this chapter thus is to present the legal background of the Hungarian local self-government system, to point out the impact of the economic crisis on Hungarian local self-government, to underline the reforms and the results. The chapter highlights the weaknesses of the local self-government's economy, and the methods which were applied in Hungary in order to fend off the negative impacts of the crises. The related funds (e.g. state subsidies and own revenues) were not sufficient to carry out the duties of local self-government; therefore, the debts increased significantly since they were formed about twenty-five years ago. Accordingly, serious financial problems occurred, which deepened significantly due to the economic crisis. This phenomenon shows that effective supervision and monitoring is needed, real local autonomy based on appropriate regulation is required, in order to guarantee the fulfilment of local tasks.

Keywords Financial crisis · Local self-government reform · Urban governance · Hungary

A. Fábián (✉)
University of Pécs, Pecs, Hungary
e-mail: fabian.adrian@ajk.pte.hu

© Springer International Publishing Switzerland 2017

C. Nunes Silva and J. Buček (eds.), *Local Government and Urban Governance in Europe*, The Urban Book Series, DOI 10.1007/978-3-319-43979-2_4

4.1 Introduction

The establishment of local self-governments in 1990 had a significant importance in the history of Hungary: the name of the state administrative system, which had been used for half a century, was changed from "state administration" to "public administration". The new structure was extended with a new subsystem of local self-government administration, new organizational principles were introduced (e.g. real decentralization and autonomy) and the importance of certain principles of operation declined (e.g. state guidance), at the same time that of others increased (e.g. the principle of legality).[1]

After the full review of Act XX of 1949 on the Constitution of the Republic of Hungary,[2] local communities gained independence and were granted the right to independently regulate and manage local public affairs in compliance with the framework established by law. Autonomy made the interests and peculiarities of the particular settlements become known as the result of a legally managed correct procedure and made possible for local self-governments to perform their tasks and exercise their authority independently (O'Toole 1994).

All these were accompanied by economic independence guaranteed by the Constitution. In the period of transition, a liberal and relatively modern system of local government institutions was developed on the basis of the provisions of the Constitution: the principles of the European Charter of Local Self-Government prevailed; democratic power could be exercised locally; the system offered scope for self-regulatory processes and local legislation (Csefkó and Pálné 1993).

The convention, adopted in Strasbourg on 15 October 1985, was announced by Act XV of 1997 on the European Charter of Local Self-Government. The Charter was created under the auspices of the Council of Europe, and its purpose was to specify standards, deriving from the rule of law and democracy, which are generally applied in nearly fifty member states of the Council of Europe in course of establishing their respective systems of local self-governments.

A certain democratic mechanism was developed, in which "centralization, which may be regarded as having a general effect, can prevail in the interest of achieving social aims, while in the interest of achieving all other aims of public interest, partial self-governance (autonomy) can prevail" (Tamás 1997). The peculiarities of the Hungarian local self-government system, developed this way, stem from several sources: Hungarian traditions of self-government, the institutions of the former Soviet-type council system, which were considered appropriate within the framework of a constitutional state, inspired in Western-European (mainly South-German) self-governmental systems. The modern structure of Hungarian self-government is based on the aforementioned facts. The structure of Hungarian

---

[1]"One of the most important legislative tasks of these months and even of this year is to adopt the act on local governments and to hold local elections", said Antall József, Prime Minister in the Hungarian Parliament on 22 May 1990.

[2]In effect until 1 January 2012, hereinafter referred to as the Constitution.

local self-government is still established on two other layers: settlement-level and county (regional level) self-government. Task performance (and financing) is focused on settlement-level self-governments. Since 1990, county self-governments have been seeking their place in the Hungarian self-government administration (Szabó 1994).

Although self-government has a double character, service and (public) authority, it is indisputable that local self-government provide certain local public services, but their organs rarely participate in the exercise of local public authority.

By the end of the first decade of the new millennium, it became obvious that the self-government system suffered from internal conflicts and was not sustainable due to the steadily decreasing state subsidies and the impact of economic downturn; furthermore, it was grievously unfair from several points of view (Kákai 2010).

The problem is a contradiction: local self-government should provide public services and functions for the local community, but because of the economic circumstances they are not able to finance, to manage, to fulfil their tasks laid down in laws. In the last twenty years, the question has been how to solve this situation, but the economic crisis accelerated the need of finding the real solution. Some authors' opinion was that the parliament should reduce the number of local self-government's tasks (Hoffman 2013), some argued for new financing system for local self-government (Dankó and Lóránt 2010). The introduced reforms used a blended conception with extraordinary elements as well (Lentner 2014).

This chapter examines the constitutional basis for local self-governments in Hungary, the tasks and functions, the relationship between local self-government and state organs, besides the related economic conditions. The aim of this chapter was to point out the impact of the crisis on Hungarian local self-government, and indirectly on urban governance processes, the kind of reforms that has been introduced and the results of the reforms.

4.2 The Constitutional Status of Local Self-Government in Hungary

The Constitution of Hungary, abrogated on 1 January 2012, deals with local self-government in a quite detailed way, so does the Fundamental Law of Hungary, which entered into force on 1 January 2012. Five articles and twenty-three paragraphs of the Fundamental Law deal with local self-governments. The territorial division of Hungary is specified in Article F) in the part titled "Foundation of the Fundamental Law".[3]

[3]The capital of Hungary is Budapest. The territory of Hungary consists of the capital, counties, cities and towns, as well as villages. The capital, as well as the cities and towns, may be divided into districts. Provisions pertaining to public authority at a local level can be found in the part called "Local self-government". The fact that "constitutional statutes", called cardinal Acts, detailing special rules pertaining to local self-government (to be adopted later) are referred to four

The Fundamental Law, unlike the provisions of the former Constitution, makes no reference to the content of local self-governance, local autonomy or the fundamental constitutional right to local self-governance, which local citizens are entitled to. Obviously, local citizens can still participate both directly and indirectly in the exercise of local power.[4]

There are minimal, hardly noticeable changes in the text compared to the previous regulation. The most important change was the title of the article: instead of the term "fundamental rights of self-government" used formerly, the above-mentioned rights are named as "the responsibilities and competencies of local self-government" in the Fundamental Law. This term—compared to fundamental rights—better matches the nature of local self-government as administrative organs (Fábián 2011).

The possibility of intervention granted to county (metropolitan) government offices is relatively far from the modern supervisory methods (e.g. consultation, notice) applied to prevent violations that local self-governments might commit. The primary goal of state supervision is to ensure the lawful operation of local self-government. State organs must facilitate the performance of local self-government while striving to assert the constitutional principle of the legality of public administration. A further goal of state supervision is to help local self-government perform their tasks by providing advice,

(Footnote 3 continued)

times in this part indicates that essential content elements of legal regulation appear in the detailed rules. ["Cardinal Acts shall be Acts, the adoption and amendment of which require the votes of two-thirds of the Members of Parliament present." Article T(4) of the Fundamental Law.] The provisions pertaining to the territorial division of the country and to local self-governments are structurally in the Fundamental Law. In Hungary, local governments shall be established to administer public affairs and exercise public power at a local level and the basic rules are to be defined by a cardinal Act [Article 31(1) of the Fundamental Law].

Article 32 of the Fundamental Law sets forth that in administering local public affairs local governments shall, to the extent permitted by law:

(a) adopt decrees;
(b) adopt decisions;
(c) perform autonomous administration;
(d) determine their regime of organisation and operation;
(e) exercise their rights as owners of local self-government properties;
(f) determine their budgets and perform independent financial management accordingly;
(g) engage in entrepreneurial activities with their assets and revenue available for the purpose, without jeopardising the performance of their compulsory tasks;
(h) decide on the types and rates of local taxes;
(i) create local self-government symbols and establish local decorations and honorary titles;
(j) ask for information, propose decisions and express their views to competent bodies;
(k) be free to associate with other local self-governments, establish alliances for the representation of interests, cooperate with the local governments of other countries within their competences and be free to affiliate with organisations of international local self-governments, and exercise further statutory responsibilities and competences.

[4]A provision in the chapter titled Freedom and responsibility lays down, "Every adult Hungarian citizen shall have the right to vote and to be voted for in elections of Members of Parliament, local self-government representatives and mayors, and Members of the European Parliament". [Article XXIII(1) of the Fundamental Law].

support and protect local communities and enhance the sense of responsibility of local self-government organs.

The responsibilities and competences of local self-government shall be exercised by local representative bodies. Local representative bodies are headed by mayors. County representative bodies elect one member to serve as president for the term of their mandate. Local representative bodies may elect committees and establish offices as defined by a cardinal Act.[5]

Local self-government and state organs shall cooperate to achieve community goals. An Act may define compulsory responsibilities and competencies for local self-government, which shall be entitled to proportionate budgetary and other financial support for the performance of their compulsory responsibilities and competences. An Act may authorize local self-governments to perform their compulsory duties through associations.

An Act or a government decree authorized by an Act may exceptionally specify duties and powers relating to public administration for mayors and presidents of county representative bodies.

Establishing the institution of mandatory local self-government associations, making it possible to provide for it by law, may serve further modernization. In the interest of effective task performance, former government practice tried to make settlement-level municipalities fulfil their tasks jointly by budgetary financial means, while in the future, by virtue of the Fundamental Law, this will also be possible under a statutory provision.

The Government shall perform the legal supervision of local self-government through the metropolitan and county government offices.[6] State control (supervision) of local self-government has been a fundamental issue of the Hungarian self-governmental system since before 2012. The multitude of remedial and control mechanisms is a peculiar feature of the Hungarian local self-government system, but at the same time, it can make the system weak and dependent. It is true that there are enough—internal and external—organs (county government offices, prosecution services, State Audit Office, self-government committees, clerks, auditors and so on) to supervise the legality of the operation of local self-governments, but these organs have insufficient corrective powers.

The fundamental purpose of the Hungarian Act on Local Self-Government, adopted on the basis of the provisions of the Fundamental Law, is to establish a modern, cost-effective and task-oriented self-government system, which allows for democratic and effective operation and at the same time—in a manner asserting and protecting the collective rights of the electorate to self-governance—imposes stricter limits on self-government autonomy than before. The provisions of the new Act on Local Self-Government allow the inclusion and operation of renewed structures (e.g. differentiated transfer of powers).

---

[5]Article 33 of the Fundamental Law.

[6]Article 34 of the Fundamental Law.

Although it is a separated subsystem, the budget of the municipalities is part of the national budget. The separation does not exclude the subsidy for local self-government to be provided by the state (by central government).

The finance of local self-government is based on the annual budget of the municipality. The funding of the mandatory and voluntary municipal tasks and the delegated administrative powers is based on this legal norm. A significant change of the new Municipal Act is that operational deficit cannot be planned, and thus, the expenditures of the performance of the municipal tasks shall not exceed the revenues. Therefore, deficit can be planned only for financing investments and developments.

Municipal tasks can be funded by own revenues, received funds and state subsidies. The Act on Local Self-Government of Hungary states that the local self-government is burdened with the consequences of loss management, the central government is not responsible for the obligations of the municipalities (Kecső 2013).

The legality of the economic decisions is supervised by the county (metropolitan) government office. The economic activities of local self-government are controlled by the State Audit Office of Hungary, which controls and monitors the legality, the expediency and the effectiveness of these decisions. The subsidies co-paid by the European Union are controlled by an independent regime.

4.3 Local Governments Tasks

The Hungarian Municipal Act of 2011 is based on the system of the former Hungarian municipal law. Thus, the municipal and the delegated administrative tasks are distinguished by the new Municipal Act as well. Municipal tasks can be classified into two groups: mandatory and optional tasks of the local self-government (Hoffman and Fábián 2014).

The possible municipal tasks of local self-government are listed by section 13(1) of the Hungarian Municipal Act. It is controversial in the Hungarian legal literature whether these tasks are imposed by this section directly or if it is only a non-exhaustive list. The grammatical interpretation of the Act shows that this list determines mandatory tasks: the text of this section refers that municipalities "shall perform" these tasks. However, for central government, the new Act just exceptionally contains direct power, because these tasks and services are determined by sectoral acts, in which the powers of central and local self-governments are defined[7] (Hoffman 2013).

[7]The Municipal Act defines the principles of granting powers (e.g. the differentiated imposition of tasks). The approach of the ministerial reasoning can be justified by the provision of section 13(1), which specifies such services as well which are defined as optional tasks by sectoral acts. Point 7 of the referred paragraph, for example, specifies the maintenance of theatres as a municipal task, which is defined as an optional municipal service by Act XCIX of 2008. Similarly, tasks related to local public safety are defined as voluntary services by section 17 of the Municipal Act.

One of the main amendments to the new Municipal Act does not contain the principle according to which local public affairs shall be performed by local self-government. Thus, the Parliament is entitled to define and re-define the share of powers between central and local self-government. With this new regulation, important public services, which had been organized formerly by local self-government, became the responsibility of central government and the de-concentrated agencies thereof. For example, primary and secondary schools, health care and residential social care are mainly organized by de-concentrated agencies of central government (Hoffman 2012).

Mandatory municipal tasks are defined by an act based on the Municipal Act.[8] The tasks of the different municipalities should be defined differently by sectorial or special regulations.[9] Thus, the nature of the duty and the different capabilities of local self-government, especially their different economic performance, population and the size of the area of the municipality shall be taken into account.[10] The personnel, the material and the financial conditions of the performance of the obligatory tasks (public services) can be regulated not only by acts but also by the decrees of the Hungarian Government and by the decrees of the ministers under these general rules of the municipal law. This right of the central government to regulate the conditions of (local) public services is not unconditional. Resolution 47/1991 (24 Sept.) of the Hungarian Constitutional Court declared that the decree, which entirely excludes the free decision of local self-government breaches the constitution. The performance of the mandatory municipal tasks has priority, because the performance of these duties cannot be jeopardized by the performance of the optional tasks of local self-government.

Due to the continental (general clause) approach of the Hungarian local self-government system, tasks which are not required by acts, namely the optional tasks of local self-government, can also be performed. The main aim of municipalities is the fulfilment of the mandatory tasks; thus local self-government can provide the optional tasks only if strict legal conditions are met. Firstly, only local public affairs may be performed as an optional task. Local self-government can perform such a task, which does not belong to the responsibility of the central government.

"Therefore, the Constitutional Court declared that a local self-government decree establishing a city police with the powers and duties of the (state) police is a breach of the constitution.[11] Secondly, the performance of optional tasks cannot be contrary to the law. As mentioned above, obligatory tasks have priority. The performance of

---

[8]According to a significant modification of the regulation, new instruments of legal supervision could guarantee the fulfilment of these tasks. In addition to the new instruments of legal supervision, the differentiated imposition of tasks is required. Although this differentiation was allowed by Act LXV of 1990 on Local Self-Governments, it is required by the New Municipal Act.

[9]The main criteria for the definition of these tasks are determined by Act CLXXXIX on the Local Self-Governments of Hungary.

[10]Section 11(2) of the New Municipal Act.

[11]Res. No. 8/1996. (23rd February) of the Constitutional Court.

optional tasks can be funded only by own revenues of the local self-government and by special central subsidies for these tasks as specified by the Act on the Budget of Hungary. Thus, Hungary has a unified state police system, where the police is maintained and directed by the central government. However, special regulations pertaining to the tasks of local public safety are determined by the Act on the Local Self-Government of Hungary. Municipalities are allowed by the new Municipal Act to establish an organization responsible for local public safety and for the preservation of local self-government assets. This organization can use force determined by acts. This task is obviously optional. Due to the use of force, the central government has stronger supervision: the (state) police has not only legal but technical supervision powers, as well. Therefore, municipalities should conclude agreements with the police" (Hoffman and Fábián 2014).

4.4 Relationship Between Local Self-Government and State Organs

Parliament shall regulate by law the legal status, powers and responsibilities, mandatory tasks, types of organs, guarantees of operation and financial resources of local self-government and the basic rules of their financial management. Parliament shall decide on the dissolution of a representative body that breaches the Fundamental Law on its first session following the receipt of the initiative of the Government.

Parliament shall decide on the territorial division of the state and—after having consulted with the concerned local self-government—the merger and separation of counties, changes in their boundaries, their names and capitals, and the formation of Budapest districts.

The Government shall, with the participation of the minister in charge of the legal supervision of local self-government, provide the legal supervision of local self-government through the government office. The minister in charge of the legal supervision of local self-government shall:

(a) run the legal supervision of local self-government;
(b) initiate the submission of a proposal by the Government to Parliament for the dissolution of any local self-government operating in violation of the Fundamental Law;
(c) prepare the decisions on territorial issues that fall within the scope of the Parliament's and the President of the Republic's jurisdiction;
(d) after having examined the proposal of the government office—provided all conditions are met—initiate the submission of a Government proposal requesting the Constitutional Court to review the constitutionality of a local self-government's decree.

The minister in charge of local self-government shall:

(a) take part in the preparation of the drafts of statutes, legal acts for the governance of public organizations and the drafts of specific state decisions, concerning powers and responsibilities of local self-government, and activities of mayors, the Lord Mayor, and the government office;
(b) take part in granting government consent to specific borrowings and commitments of local self-government;
(c) coordinate the Government's responsibilities concerning issues of settlement development and financial management connected to the operation of local self-government and issue of certain databases;
(d) coordinate the tenders for the utilization of development funds made available to local self-government and the establishment of the legal conditions concerning the financial management of local self-government;
(e) take part in collecting and systemizing local government data, necessary for the system of task financing of local self-governments;
(f) be entitled to institute, through the minister in charge of the national budget, an on-the-spot audit of the financial management of local self-government performed by the Hungarian State Treasury;
(g) be entitled to have access to and systematize all the data of the information system of public finances concerning local self-government subsystem of public finances.

National municipal alliances may put forward proposals concerning legislative tasks and measures to take and may consult with the Government about strategic issues concerning public services provided by local self-government and issues of the national budget concerning local self-government (Hoffman and Fábián 2014).

Beyond the competences defined in the Fundamental Law, in the sphere of legal supervision over local self-government,[12] the government office may:

(a) issue a legal notice;
(b) initiate the convocation of the representative body or the association council and convene the representative body or the association council in cases defined by law;
(c) propose the minister in charge of local self-government to initiate the submission of a Government proposal requesting the Constitutional Court to review the constitutionality of a local self-government decree;
(d) initiate the review of a local self-government decision at an administrative court;

[12]In a legal supervision procedure, the government office shall examine the legality of the organisation, operation and decision-making procedure of the local self-government; the legality of the decisions (decrees and decisions) of the local self-government; the fulfilment of the local self-government's fulfilment of their statutory legislative obligation, and the fulfilment of their decision-making or task performance (public service provision) obligation.

(e) initiate the commencement of a court proceeding against a representative body for failure of the decision-making or task performance obligation and for ordering substitute decision-making;
(f) propose the minister in charge of the legal supervision of local self-government to initiate the submission of a proposal by the Government for the dissolution of any local self-government breaching the Fundamental Law;
(g) initiate at the Hungarian State Treasury the withholding or withdrawal of a specific part of a subsidy, defined in law, due from the national budget;
(h) start proceedings against a mayor having committed repeated infringements for terminating his mandate;
(i) initiate a disciplinary procedure against the mayor of a local self-government or against a town clerk;
j) initiate the audit of the financial management of a local self-government by the State Audit Office;
(k) provide professional help for local self-government in matters falling within the scope of their powers and responsibilities;
(l) impose a legal supervision fines on local self-government or associations in cases defined in Parliament's Act.

The government office, in respect of matters falling within the competence of the local self-government, may request information or initiate consultation, which must be performed by those concerned within the deadline set (Hoffman and Fábián 2014).

In case the government office notices a violation of law, within the scope of legal supervision, it shall issue a legal notice (not shorter than 30 days) on those concerned to eliminate the violation within a specified deadline. Those concerned shall examine the contents of the legal notice and notify the government office about the measures taken accordingly or about their disagreement within the deadline set. If the deadline is not met, the government office decides on the application of other means of legal supervision procedure within the scope of its discretionary power.

If the government office finds that a local self-government decree breaches the Fundamental Law, it may propose the minister in charge of the legal supervision of local self-government to request the Constitutional Court to review the constitutionality of the local government's decree. The minister in charge, during the legal supervision of local self-government, after having examined the proposal of the government office—provided all conditions are met—shall initiate the submission of a Government proposal requesting a review as to whether the local government decree complies with the Fundamental Law.

The government office may request the Curia (High Court of Hungary) to review whether the local self-government decree complies with statutory provisions within fifteen days reckoned from the receipt of the notification from the local self-government or the lapse of the deadline. Upon the commencement of the court proceedings, the government office shall deliver the motion to the concerned local self-government.

"The government office shall request the Curia to establish the failure of the local self-government to perform its legislative obligation if the local self-government has failed to perform its statutory obligation to legislate, and shall simultaneously inform the local self-government about it. If the local self-government fails to fulfil the legislative obligation thereof within the deadline set by the Curia, the government office, within 30 days reckoned from the lapse of the deadline, shall request the Curia to order the government office to remedy the omission" (Hoffman and Fábián 2014).

The decree adopted by the head of the government office in the name of a local self-government shall qualify as a local self-government decree with the exception that the local self-government is entitled to amend or repeal it only after the next municipal elections. During this period, any amendments may be made by the head of the government office.

Within fifteen days reckoned from the receipt of the notification from the local government or the lapse of the deadline for the notification, the government office may request the administrative court to review the local self-government decision. If the implementation of the local self-government decision found to be in violation of law would cause a serious injury to the public interest or would cause unavoidable damage, the court shall suspend the implementation of the decision.

The government office may impose a legal supervision fine on local self-government. The legal supervision fine may be imposed repeatedly if the same obligation is breached repeatedly or in case of the breach of any other obligation. When imposing the legal supervision fine, the government office shall take into consideration:

(a) the seriousness of the breach of obligation;
(b) the budgetary situation of the local self-government; and
(c) the number and amount of fines imposed previously.

The judicial review of the decision of the government office imposing a legal supervision fine may be requested from the competent administrative court by the local self-government or association within fifteen days from the receipt of the decision (Hoffman and Fábián 2014).

4.5 The Impact of the Crisis on the Economic Position of Hungarian Local Self-Governments

The economy of the Hungarian local self-government was unstable in the years preceding the crisis, they were unable to carry out their duties, self-government debts continued to grow. Their incomes and the value of their revenues have decreased since 2008, the sum of their loans continued to grow, and the loans financed mostly their spending and not economic development, investments or reserves. Hungarian local self-governments tried to issue bonds also. The end result was that the amount of the obligations of Hungarian municipalities increased six-fold to 3 billion Euros between 2002 and 2008.

The problem was enhanced by the fact that a large part of local self-governments' obligations was in foreign currency (mainly in CHF), and the rise in the currency exchange rates increased the amount of the debts (Fig. 4.1). The Hungarian government has attempted to apply multiple devices for solving the financial crisis of Hungarian local self-government since 2010: decreasing the number of municipal functions, adopting more stringent rules for assuming new contractual obligations, carrying out debt consolidation, introducing new forms of financing and enhancing the supervision of managing self-government economy.

After 2007, the economic situation of Hungarian local self-government began to deteriorate significantly (Halmosi 2013). The crisis hit hard local self-governments' management performance and their financial capacity as well. As a consequence, the external indebtedness of the sector increased dramatically in recent years.

The debt increase was caused by several factors. First, local self-government cannot continually rely on collecting their own revenues based on their everyday operation since local taxes are collected only twice a year (in March and in September), according to the corresponding legal provisions. Second, the lack of own resources for launching bigger investments was also responsible for the debt increase. Third, the subsidies provided by the state were frozen, which was one of the reasons for local self-government to use external funds, in order to finance their current operations and investment expenditures. Therefore, bonds were issued, and/or bank loans were obtained (Ercsey et al. 2013).

The marked indebtedness of the local self-government sector began in 2006 primarily in the form of municipal bonds (Fig. 4.2).

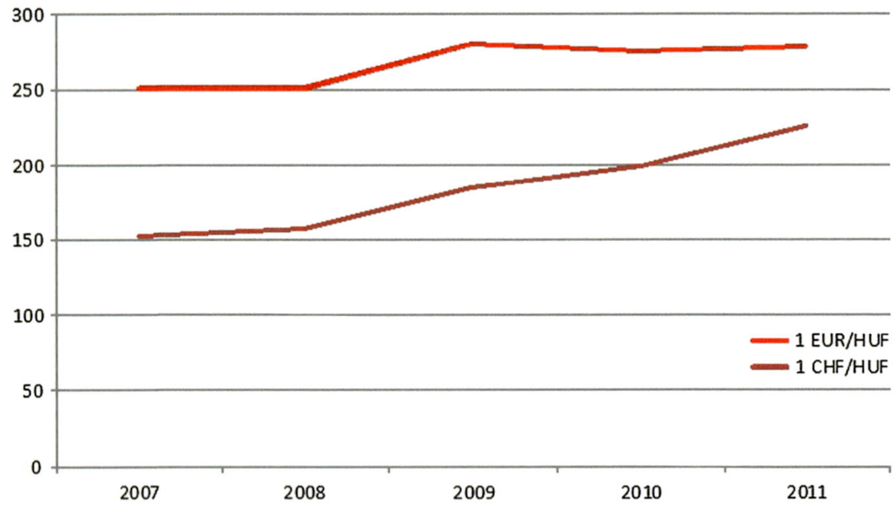

Fig. 4.1 Changes of EUR/HUF and CHF/HUF exchange rates (yearly average). *Source* Hungarian National Bank

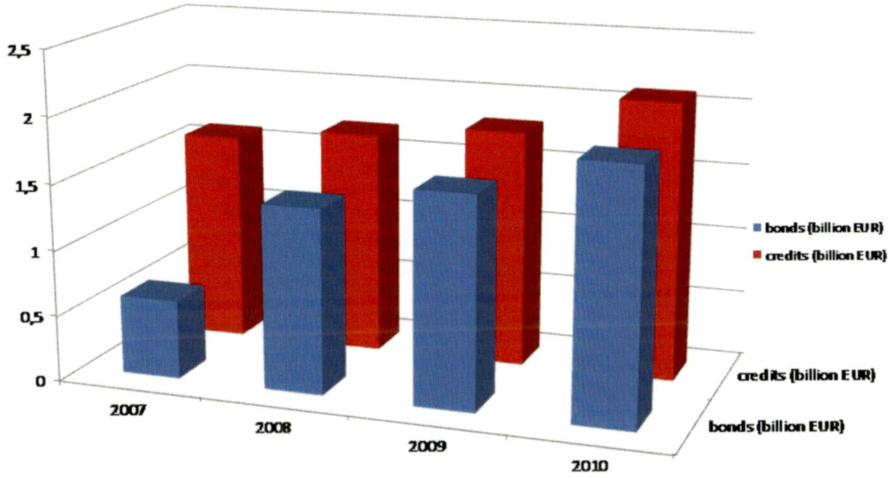

Fig. 4.2 Changes of Hungarian local self-government debts, billion EUR, 2007–2010. *Source* Mohl (2012)

In 2006, bonds of approximately 8 billion HUF (total nominal value) were issued; in 2007, this sum was similar, while in 2008, it exceeded 200 billion HUF (approx. 0.7 billion EUR). As a result, by the end of 2008, the HUF denominated bonds increased to almost 50 billion HUF, bonds denominated in a foreign currency (CHF, EUR) rose to over the amount of 350 billion HUF (approx. 1.2 billion EUR) (Kovács 2010).

In Hungary, the so-called debt consolidation was a very important tool for the management of local self-governments' debt.[13]

The Hungarian state took over 100 % of the debt in the case of local self-government with less than 5000 inhabitants. In the case of local self-government with more than 5000 inhabitants, the rate of the consolidation varied between 40 and 70 %. First, the consolidation of municipalities with a population under 5000 was implemented. The state paid back the total debt of 1710 settlements (with less than 5000 inhabitants) in the amount of 74 billion HUF, which came from 3848 contracts assumed. From the debt of settlements with a population over 5000, the Hungarian state took over 477 billion HUF (approx. 1.6 billion EUR).

The last phase of consolidating municipalities took place in the spring of 2014. The remaining debts of municipalities with more than 5 thousand inhabitants have been taken over by the state. The total cost assumed ultimately amounted to 1344 billion HUF (approx. 4.5 billion EUR) (Lentner 2014).

The borrowing of local self-government tends to serve the purpose of operation and the performance of compulsory tasks instead of financing investments and

[13]Legal background: Act on State Budget of 2012, section 76/C; Act on State Budget of 2013, sections 72–75.

developments. According to Article 34 of the Fundamental Law, an act may stipulate the conditions for, or require the Government's consent to, any borrowing above the statutory limit or any other commitment of local self-government with the aim of preserving their budget balance. Obviously, the deficit in the budget of local self-government is caused typically by the insufficiency of state subsidies and own revenues to cover the expenses of performing compulsory tasks and providing local public services.

The Fundamental Law cannot solve the issue of financing, only the tool for "a debt brake" has been institutionalized. The effectiveness thereof is heavily disputed, and it severely restricts local economic autonomy. It should also be added that the effectiveness of this provision is further endangered by the delay thereof: credit institutions—noticing the enormous problems of managing the property and financing of local self-government—tend to be less willing to finance the operation of local self-government regardless of whether the Government will consent borrowing or not.

The local self-government finance is based on the annual budget of the municipality. The funding of the mandatory and voluntary municipal tasks and the tasks of the delegated administrative power is based on this legal norm. A significant change in the new Municipal Act is that operational deficit cannot be planned, and thus, municipal expenditures shall not exceed the revenues. Therefore, deficit can be planned only for financing investments and developments.

Municipal tasks can be funded by own revenues, external funds and state subsidies. The Act on the Local Self-Governments of Hungary states that the local self-government is burdened with the consequences of loss management, and the central government is not responsible for the obligations of the municipalities (Kecső 2013).

The following public revenues are considered municipal own revenues: fees and charges of municipal services and of municipal asset management, dividends, profit of the municipal business activity, rent, funds received as private incomes of local self-government and local taxes, fees and fines (Ercsey 2015). Local taxes include local business tax, tourism tax, communal personal income tax, land tax and building tax.

The main changes of the regulation, pertaining to own revenues, are the new limitations imposed on municipal loans. As mentioned above, the requirement of Government approval for local self-government borrowing was introduced by the Article 34(5) of the Fundamental Law. The aim of this provision is to prevent local self-governments' indebtedness.[14]

In principle, all loans and other transactions with a nature of loan (e.g. municipal bonds) shall be permitted by Central Government. There are broad exceptions to this principle. Similarly, loans which are required for the financing of projects with the co-financing of the European Union do not need the consent of the Hungarian

[14]Detailed rules are prescribed by Act CXCIV of 2011 on the Economic Stability of Hungary.

Government. Although there are a huge number of exceptional cases, the financial freedom of local self-government is significantly limited by this legal institution.

The assigned central taxes count as revenues of local self-government, but their significance is weak. One example of an assigned central tax is the 40 % of the vehicle tax (a tax collected by local tax authorities): Another example is the 100 % of the personal income tax on land rent.

The rules on state subsidies were significantly changed by the new Municipal Act. In 2013, a task-based financing system was introduced. Thus, state subsidies are based on the mandatory tasks of municipalities. Firstly, they depend on the standards of services defined by legal norms. The efficient management, the expected own revenue of the municipality and the actual revenues of local self-government have to be taken into account when determining subsidies.

The main principle of the task-based financing system is its additional nature: the own revenues of local self-government are complemented by the state subsidies, and thus, local communities are interested in collecting their own revenues (Kecső 2013). The task-based subsidies are earmarked; thus, the expenditure shall be spent on the financing of the mandatory tasks defined by the act on the annual central budget and on certain voluntary tasks defined by the budget of the municipalities.

Local self-government is responsible for their economic management; thus, local self-government can also go bankrupt. The procedure of discharging bankrupt municipalities from debts is regulated by Act XXV of 1996.

First of all, the legality of economic decisions is supervised by the county (metropolitan) government office. The economic activities of local self-government are controlled by the State Audit Office of Hungary, which controls and monitors the legality, expediency and effectiveness of these decisions. The subsidies co-financed by the European Union are controlled by an independent regime.

"Economic control and monitoring within the local self-government system were amended partially. The monitoring powers of the economic committee of the representative body were extended. Similarly to the former regulation, the internal control is performed by the clerk. The internal audit was simplified by the new Municipal Act because an audit by independent auditor companies is no longer required by the municipal law" (Hoffman and Fábián 2014).

4.6 Conclusion

The legal background in Hungary is suitable for the operation of local self-governments to fulfil the local tasks. The financial problems of local self-government after the financial crisis were not caused by insufficiency in the legal regulation. The main reason existed already before the crisis: the underfunding of local self-government, and the local self-governments' answer to this situation. It follows that the modification of the regulation could not solve the economic problems of the Hungarian local self-government. The debt consolidation can be considered as a final solution, but the underfunding could be eliminated only if there is a proportionate financial state support for providing local public services. This is stated in the European Charter of Local Self-Government and in the Fundamental Law of Hungary[15] as well.

The Report of Monitoring Committee on Local and Regional Democracy in Hungary[16] recommends that the Committee of Ministers invite the Hungarian authorities to grant local authorities financial autonomy, in order to enable them to exercise their powers properly, in particular by adjusting the level of grants allocated by the central government to local authorities so that their resources remain commensurate with their powers and by limiting central government supervision of the management of local finance so that it is proportionate within the meaning of Article 8 of the Charter. In other words, as Dankó and Lóránt (2010) refer "the lack of harmony between the tasks and resources of local self-governments, partly due to cutbacks in per capita supports and other subsidies and also for the lack of a comprehensive revision of the range of tasks, authority and financing, has been reproduced every year".

The financial background of local self-government was restricted, and the protection of assets was strengthened by the new regulation. These changes have been justified by the prevention of local self-government indebtedness and by more efficient national asset management.

The challenge of the coming years for the Hungarian government might be to find a balance between the cost of providing local public services and the amount of financial state support, a process that will certainly have impacts on urban governance. Strengths and weaknesses of the new system need further research. It must be highlighted that an effective supervision over local self-government without proper legislation, without real local autonomy, without stable economic background is not able to ensure the local functions.

The economic crisis has had a considerable negative effect on the financing possibilities of Hungarian local self-government. The dramatic increase in the volume of local self-government debt started in 2006 and was primarily caused by the issue of local self-government bonds. Similar patterns can be identified when examining the recent practice of Hungarian local self-government bond financing. The regulation on credit financing of Hungarian local self-government is still primarily based on the coercive effect of credit limit, which is unjustified and one-sided.

---

[15]Local governments and state organs shall cooperate to achieve community goals. An Act may set out mandatory functions and powers for local governments. For the performance of their mandatory functions and powers, local governments shall be entitled to proportionate budgetary and/or other financial support. Article 34(1) of the Fundamental Law.

[16]31 October, 2013.

References

Csefkó F, Pálné K (eds) (1993) Tények és vélemények a helyi önkormányzatokról. MTA RKK, Pécs

Dankó G, Lóránt Z (2010) Challenges of the local government regulation. Public Finance Q 55 (3):511–517

Ercsey Zs (2015) Tax reforms in Hungary. Harthmedia Ltd., Pécs

Ercsey Zs, Jerković E, Perić R, Szilovics CS (2013) The finances of local and regional self-government units. In: Drinóczi T, Župan M (eds) Law-regions-development. PTE, Pécs, pp 221–254

Fábián A (ed) (2011) 20 éves a magyar önkormányzati rendszer. Jövő K A, Pécs

Halmosi P (2013) A gazdasági válság hatása az OECD országok önkormányzataira. Pénzü Szle 58 (3):296–309

Hoffman I, Fábián A (2014) Local self-governments. In: Patyi A, Rixer Á (eds) Hungarian public administration and administrative law. Schenk, Passau, pp 320–350

Hoffman I (2012) Some thoughts on the system of tasks of the local autonomies related to the organisation of personal social care. Lex Localis 6(31):323

Hoffman I (2013) Az önkormányzati feladat- és hatáskörök átalakított rendszere az új önkormányzati törvényben. Új Magyar Közigazgatás 6(7–8):10

Kákai L (ed) (2010) 20 évesek az önkormányzatok Születésnap, vagy halotti tor?. PTE BTK, Pécs

Kecső G (2013) A helyi önkormányzatok gazdálkodásának reformja és az adósságrendezési eljárás módosításai. Új Magyar Közigazgatás 6(7–8):26

Kovács G (2010) A magyarországi helyi önkormányzatok hitelfinanszírozása a válság tükrében. In: Losoncz M, Szigeti C (eds) Válság közben fellendülés előtt. file:///C:/Users/user/Downloads/%C3%B6nk_gazd_v%C3%A1ls%C3%A1g/kovacs_gabor.pdf. Accessed 18 Jan 2015

Lentner C (2014) A magyar önkormányzatok adósságkonszolidációja. Pénzügyi Szemle 59 (3):330–344

Mohl A (2012) Az önkormányzati adóssághoz (hitelekhez, kötvényekhez) kapcsolódó finanszírozási kockázatok. http://www.asz.hu/eloadasok/2012-04-17/az-onkormanyzati-adossaghoz-hitelekhez-kotvenyekhez-kapcsolodo-finanszirozasi-kockazatok/mohl.pdf. Accessed 23 Jan 2015

O'Toole LJ (1994) Local public administrative challenges in post-socialist Hungary. Int Rev Adm Sci 60(2):293

Szabó L (1994) A középszint az önkormányzati törvény módosításának tükrében. Magyar Közigazgatás 44(12):721

Tamás A (1997) A közigazgatási jog elmélete. Szent István Társulat, Budapest

Chapter 5
Local Government Innovation in Italy and its Impact on Urban and Regional Planning with a Focus on the Milanese Context

Renzo Riboldazzi

Abstract This chapter focuses on the local government reform in Italy, with a particular attention to the regional governance and planning innovations. It aims to carry out some reflections on the contents and approach of the reform and on some of its foreseeable impact, also through the analysis of the recent debate in Italy. The Milanese context is assumed as a case study. Here, as in the rest of Italy, the application of the reform is still in process and its effects are difficult to assess, namely those that it is likely to have on the institutional framework of urban governance in the country. Nonetheless, we can see that—for many aspects and despite the efforts made by the relevant bodies to its application—it appears to have several limits. It is likewise clear that the opportunity offered by this reform should not be wasted.

Keywords Local self-government reform · Regional governance · Planning innovations · Urban governance · Italy

5.1 Introduction

Law n. 56 of 7 April 2014—called *Disposizioni sulle Città metropolitane, sulle Province, sulle unioni e fusioni di Comuni* (Provisions on Metropolitan Cities, Provinces, and the Union and Merging of Municipalities)—has paved the way for a detailed reorganisation of Italy's local government entities, which is still ongoing, and through that indirectly on urban governance. It represents the newest legal measure in this area, with the preceding general reform dating back 24 years—Law n. 142 of 8 June 1990: *Ordinamento delle autonomie locali* (Organisation of Local Self-government). Over the years, this was followed by various partial yet key measures relating to the development of regional governance models. These

R. Riboldazzi (✉)
Politecnico di Milano, Milan, Italy
e-mail: renzo.riboldazzi@polimi.it

© Springer International Publishing Switzerland 2017

C. Nunes Silva and J. Buček (eds.), *Local Government and Urban Governance in Europe*, The Urban Book Series, DOI 10.1007/978-3-319-43979-2_5

included Legislative Decree n. 267 on the organisation of local bodies issued in 2000; the 2001 reform of title V of the second part of the Constitution of the Republic of Italy (Constitutional Law n. 3 of 18 October 2001); Law n. 135 of 7 August 2012, which in many ways foreshadowed the recently enacted reform, yet which was stricken the year after its approval for being unconstitutional (Constitutional Court Sentence n. 220/2013) precisely because of the way it reordered local government entities; and finally, a new Government Decree for another constitutional reform—quite a bit broader than the one from 2001—with ratification discussions currently underway in Parliament (Tortorella and Allulli 2014; Vesperini 2014; Perulli 2014; Chiacchio 2014; Onida 2015).

In Italy, the creation of the Metropolitan Cities and more generally the reform of local government are "long overdue and [have come] with a law that doesn't hold water" (Consonni 2014d). In fact, some consider Law 56/2014, not unlike other Italian laws that are plagued with interpretation problems, as "rushed and confused" (Pompilio 2015: 8), "full of contradictions and intrinsic political flaws" (Beltrami Gadola 2014: 2), "null and void due to grave violations of citizen's rights and equal dignity of the Municipalities in the region", and even in violation of the 1985 European Charter of local self-government, according to the most severe critics (Natale 2014a: 5). To this is added that, "for various reasons, the transition from the old to the new regime has not been a painless process" (Rapicavoli 2015). Among these reasons is that "with the taking effect of Law 56/2014, the regional laws on regional government have become largely obsolete and require updating" (Pompilio 2014: 7). Finally, the practical application of Law 56/2014 has at times been experienced by local administrators as "just one more requirement, a bureaucratic nuisance, [and not an opportunity to] explore a new large scale policy that affects housing, labour, and the region" (Boeri 2015). Therefore, so far it has been "managed without vision, enthusiasm, participation, or courage" (Vitale 2015), to such a degree that some assert that at least with Milan, it is a "risky and dangerous front" that will fail to lead "to simplification but instead to over complication" (Ballabio 2015: 7).

This chapter aims to clearly outline the essential contents of Law 56/2014, placing particular attention on its urban planning aspects. In particular, it briefly explains the principles on which local government is founded in Italy today. It describes the administrative bodies (with specific reference to the Metropolitan Cities) and their institutional functions, including the tools that the new law provides to ensure local self-government and to support regional planning. Finally, looks at how the law is being applied in the Milanese context that is assumed as a case study.

The main research questions are the following: Is the Law 56/2014 a good reform from the urban governance and spatial planning point of view? Will this reform carry out improvements or—as many critics state—worse conditions in the Italian regional planning? What are the foreseeable critical issues of the reform concerning spatial planning?

It is to be pointed out that today a definitive answer is not possible because the application of the reform is still in process and its effects are difficult to assess. In fact, this chapter was completed in October 2015, after few months of the law coming into force. Nonetheless, also through the analysis of the current debate at that time in Italy, it is possible to formulate some hypotheses to reach the first conclusions about the effectiveness of the reform and its potential impact on urban governance.

5.2 The Founding Principles for the Reform of Local Government in Italy

Law 56/2014 has redrawn the scope and competencies of local government in Italy based on three fundamental principles: subsidiarity, differentiation, and appropriateness (section 1, paragraph 1). These principles were only introduced into the Italian Constitution as recently as 2001, a bit late compared to Germany, the USA, and the European Union—at least with regard to subsidiarity—which the latter makes specific reference to in the 1992 Maastricht Treaty. In political administrative language, the principle of subsidiarity expresses "the concept by which a central authority has an essentially subsidiary function, being assigned only those functions that the local government is not able to carry out by itself" (Treccani, Vocabolario). Differentiation is a more complex concept with various interpretations. In social systems, differentiation is commonly considered the characteristic by which systems internally "tend to develop at least partially autonomous subsystems, distinct in their segmentation (non-hierarchical distinction), stratification (hierarchical distinction), or function (division of the overall activity into multiple specialised sectors representing different components)" (Treccani, Enciclopedia). It can therefore be asserted that, at least in terms of its principles, Law 56/2014 is in keeping with section 5 of the Italian Constitution, which "recognises and promotes local government [and] implements the broadest decentralisation of government for State services". The new law seeks to transfer certain functions to more peripheral entities and, at the same time, affirm the independence of local government. The third principle, appropriateness, simply indicates the conditions of being "appropriate", i.e., "proportional, suitable, and correct" (Treccani, Vocabolario). This indicates the legislator's intention of putting measures in place to make the government apparatus harmonious and balanced in relation to the functions it has to carry out, eliminating any overlapping responsibilities. Unfortunately, the Italian system is rife with overlaps that often slow things down and increase costs to such a degree that they significantly impede the economic and social activities in the region.

5.3 Local Government Entities and Their Functions

As indicated by the title and explained in the first section, Law 56/2014 regulates three types of local government entities: the Metropolitan Cities, a new government body in Italy, as well as one of the main inspirations for the new law, that was introduced—though not yet implemented—into the Italian Constitution in 2001 (title V, part II, section 114); the Provinces, which up until the enactment of the new law were the only local entity that served as an intermediary between the Municipalities and the Regions; and finally, the Unions and Mergers of Municipalities.

5.3.1 Unions and Mergers of Municipalities (as Well as Their Partitions)

For the Municipalities, Law 56/2014 calls for two types of measures: centripetal, i.e. aimed at the concentration of administrative functions and responsibilities, and centrifugal, i.e. aimed at breaking them up. The former applies to the smaller municipalities, which is a typical Italian phenomenon. The latter applies to the capital cities of the Metropolitan Cities, which will be discussed in more detail in the next section.

The Unions and Mergers of Municipalities are two distinct measures designed to maximise efficiencies and streamline public spending, both in terms of the government apparatus and in terms of the delivery of government services. Unions will essentially allow municipalities of a modest scale to manage functions and services through association with other municipalities, yet without giving up their identity as an independent government entity. It is a measure that already existed in Italy before the implementation of Law 56/2014, so much so that the operating Unions are now 448, of which over 60 in Lombardy, for a total of 2387 Municipalities involved at a national level and an almost ten-million-inhabitant population (ANCI). Mergers of two or more contiguous Municipalities will revoke their respective individual administrative entities and create—by means of a regional law—a new, larger Municipality. Apart from special cases, such as that of the "incorporation in a contiguous Municipality", and the Regional incentive types, both measures are to be mainly carried out on a voluntary basis and—especially concerning the second—they could find limited application, as has been the case with other regulatory measures in the same vein. For example, more than 70 years ago a key urbanism law (Law n. 1150 of 17 August 1942) that called for a similar kind of inter-municipal planning didn't lead to the expected results at a national level, except for the more cultural than operating experience of Piano Intercomunale Milanese that, always in the Milanese area, was the basis of a first "attempt to establish a metropolitan government" (Balducci 2015: 41).

However, quite unlike these precedents, Law 56/2014 stipulates that, "so as to enable the holding of a popular election of a mayor and metropolitan council, by the date of the election the region of the capital city should have been split into numerous municipalities" (section 1, paragraph 22) or divided into several administrative parts. This regulation is clearly aimed at rebalancing the political weight of the municipalities included in a metropolitan area by reducing the bargaining power of the larger ones. For example, Milan has more than 1.3 million inhabitants, yet of the 133 municipalities in the Milan metropolitan area, one has less than one thousand inhabitants, six have less than two thousand, three have less than three thousand, 11 have less than four thousand, 13 have less than five thousand, while the municipalities that have five to ten thousand inhabitants are 41, those that have twenty to thirty thousand inhabitants are 14, those that have thirty to forty thousand inhabitants are 8, those that have forty to sixty thousand inhabitants are 5, only one has between 70 and 80 thousand (Cinisello Balsamo), and only one has more than 80 thousand (Sesto San Giovanni) (ANCI).

Apart from the contradictions, the Law dictated in its application—such as a referendum involving all citizens of the Metropolitan City and not exclusively, as it should be, those of the Municipality affected by the measure—should the government of a large municipality be divided "without a strategic plan [that thoroughly acknowledges its unique regional character, the result would be] total institutional butchery. Problems would not be solved, [but] rather cost and bureaucracy increased […]. It would mean just a smaller version—and in some cases, the multiplication—of the same imbalances and dysfunction that the process was supposed to address in the first place" (Consonni 2014b).

Perhaps it would be more useful and logical if the two measures—centripetal and centrifugal—were applied jointly, especially in contexts where there was an actual need. One thinks, for example, of the Milan suburbs and of the bordering municipalities. Being artificially separated by government boundaries that have nothing to do with the urban dynamics at play, these would not be "annexed […] but rather connected to [those] parts of Milan that [would] otherwise [be] made twice as peripheral" (Sarfatti 2014: 8). Therefore, it would be logical "to overcome, on the one hand, the atomisation of municipalities and, on the other, the monolithic quality of the capital cities [to] create balanced municipal systems within the metropolitan areas" (Natale 2014b: 10) without negating their historical identity or present-day conditions. More generally, this has to do with avoiding two governance models "both of which are absurd and undemocratic" (Consonni 2014c: 9), yet which are nonetheless allowed under the law. In one case, "the capital city assumes the de facto role as the governing body for the entire metropolitan area", and in the other, "the capital city disappears, fragmented into multiple municipalities, and the popularly elected metropolitan government is built on a base that bears some semblance of egalitarianism" (Consonni 2014c: 9).

This is the direction in which Law 56/2014 could be going; one which, within the context of the Metropolitan Cities, allows for the formation of "homogeneous zones for specific functions, taking into account regional specificity" (section 1, paragraph 11). It is an opportunity that the Metropolitan City of Milan seized—as it was an essential condition to re-establish the one-man-one-vote elections of its government organs—and that has been translated into the articulation of its territory into seven parts “according to geographic, demographic, historic, economic, and institutional characteristics”. Starting in September 2015, the metropolitan area of Milan has been structured into seven homogeneous zones: Alto Milanese (22 municipalities, 258,743 inhabitants, 21,523 ha); Magentino-Abbiatense (29 municipalities, 213,745 inhabitants, 36,044 ha); Sud Ovest (16 municipalities, 238,729 inhabitants, 17,994 ha); Sud Est (15 municipalities, 173,267 inhabitants, 17,972 ha); Adda Martesana (28 municipalities, 336,284 inhabitants, 26,495 ha); Nord Ovest (16 municipalities, 315,749 inhabitants, 13,582 ha); and Nord Milano (7 municipalities, 315,494 inhabitants, 5788 ha) (CMM-CM 2015a, b). On the other hand, it is unclear —especially in terms of the “identification of the activities and metropolitan services that can be decentralised [...] with the goal of promoting the integration of analogous services among individual or associated Municipalities” (SCMM, section 29, paragraph 2)—what role the Municipality of Milan will play, since it remains an autonomous entity—though it is organised into nine government areas (CM)—with no redefinition of its own government structure (CM-PM 2013).

5.3.2 *Provinces: Dying Bodies*

The second regional entity affected by Law 56/2014 is that of the Provinces. In the protracted debate that has been going on for years in Italy over the reform of local government entities, which are generally considered costly and inadequate to respond efficiently to various social, economic, and regional issues, “the [fiercest] criticisms are focused on [precisely this level of local government, the] Provinces, to the point of making them (false) sacrificial lambs” (Consonni 2014a: 3). False because in reality, with the approval of Law 56/2014, the idea has gone from abolishing them completely—an idea that still permeates the government decree of constitutional reform under discussion in Parliament—to preserving them almost entirely, perhaps scaling them down in terms of function, apparatus, and workforce. Today, the Provinces remain responsible for the schools, roads, and environmental protection. Most importantly, the Provinces no longer have direct elections for the representatives who sit on their governing bodies. In fact—if we exclude the aforementioned possibility regarding the Metropolitan Cities—starting in 2014, Italian citizens are only called upon to elect their representatives at the municipal and regional levels, i.e. the other two levels of local government. This is a decision that poses many issues related to how undemocratic local government is, with significant impacts on urban design, regional planning, and their political reason.[1] Regional coordination, which up until now had been carried out at a higher level—by the Provinces, tasked with drafting, approving, and applying the Regional Land

[1]On this theme, see: Mazza (2015).

Use Plan, will now pass to a non-elected entity, where the choices will be subject to bargaining among the mayors of the municipalities concerned. While, on the one hand, they will not likely have the same bargaining power, on the other, they will find themselves operating within the confines of "authorities or procedures other than their own reference entity", i.e. the one for which they were elected (Vesperini 2014: 792).

5.3.3 Metropolitan Cities

The last level of local government regulated by Law 56/2014 is the Metropolitan Cities. These are not a new concept since Law 142/1990 (though not implemented) had already called for their creation (Tortorella and Allulli 2014; Vesperini 2014; Onida 2015). However, they are definitely the most important feature of the new reform and nevertheless, it has been noted that, as designed, they "in large part resemble the Provinces, which are already institutionally weak and [as we have seen] are made even weaker" by this recent measure (Camagni 2014a).

5.3.4 Scope

Metropolitan Cities were intended for vast regional areas. They were not merely created to unify and encourage interaction among Municipalities of a given catchment area but also as a way to find solutions to problems that extend beyond simple government boundaries. In reality, their government sphere has not been defined based on careful consideration of the current dynamics in the relative regional catchment areas. In some cases, they simply replace the provincial government without redefining their boundaries in any way. Apart from Roma Capitale—that is included in a special law—the Metropolitan Cities, as established by Law 56/2014, are nine in the ordinary statute Regions (Milan, Naples, Turin, Bari, Florence, Bologna, Genoa, Venice, and Reggio Calabria) while four Metropolitan Cities were independently established in the special statute Regions (Catania, Messina, Palermo, and Cagliari).

The decision to have the land areas of the Metropolitan Cities coincide with the existing Provinces (section 6, paragraph 2) is considered a drawback (Roma 2014). Even if Municipalities belonging to the same Region are allowed to belong to the same Metropolitan City, in its current application, the law fails to seize the opportunity to overcome the gap so often found between the local government and the economic, social, and culture trends at work in the region. For example, take the deep economic ties between Milan and Brianza or other Lombardy areas,[2] which

[2]On this theme, see: Consonni and Tonon (2001).

are in no way represented; in fact, this geographic area is completely excluded from the new Metropolitan City of Milan, giving us a clear picture of how anachronistic and nonsensical certain regional configurations are.[3]

Rather than getting entrenched in pre-established administrative boundaries, local government reform aimed at promoting economic and administrative efficiency could have tried to overcome these pitfalls by encouraging cooperation among the Municipalities, not just in terms of land areas, but also of specific issues, services, and policies. This would have allowed for variable geometries based on the actual administrative or operating needs. Law 142/1990 "established that the Region could define the [Metropolitan City] area by reviewing the Province boundaries". Furthermore, the Metropolitan Cities would be attributed the functions of the Provinces, and those of the Municipalities that had an intra-municipal extent or for reasons of cost-effectiveness should be carried out on a coordinated basis (Onida 2015). On the other hand, this key aspect disappeared in Law 56/2014. That is apparently anachronistic if the regional settlement trends—both those of population and those of manufacturing activity—of the latest ten-year periods are considered along with men and goods flows to and from Milan. To sum up, all this underlines that "in order to understand now what Milan's urban region includes we need to consider at least ten provinces, two of which—Novara and Piacenza—belong to other regions—Piedmont and Emilia Romagna—though predominantly linked to Milan" (Balducci 2015: 44).[4]

5.3.5 *The Government Entities*

Metropolitan Cities have replaced Provinces since 1 January 2015, differing from them, albeit with the above limitations, in the type and number of functions attributed, as well as in the potential, subject to the conditions we have already mentioned and to a proper State law that still has not been passed, to restore the practice of direct elections of entity representatives. These are entities that govern areas that are economically and demographically important. "One Italian in three lives in one of the 14 Metropolitan Cities [currently instituted in Italy], which are [the areas] experiencing the highest demographic growth [and those that produce] over one third of the national GDP" (Soldavini 2014: 24). These areas are governed by the Metropolitan Mayor, who "represents the body, calls and chairs the Metropolitan Council and the Metropolitan Conference, supervises services and departments and execution of deeds" (section 1, paragraph 8). The Metropolitan Mayor is by right the mayor of the capital city (section 1, paragraph 19). This decision gives rise to a paradoxical situation: the Metropolitan City of Milan, for

[3]On the identification principles for the metropolitan areas, see: Bartaletti (2009).

[4]To better understand the deep reasons leading to this situation, see: Consonni and Tonon (2001); to reflect on possible Milanese governance models, see among others: Lanzani (2004).

instance, is governed by the mayor elected by the citizens of just one of the 134 municipalities included in it. Even this case entails a serious damage for democratic representation. The Metropolitan Mayor is not backed by executive bodies supporting the activity and sharing responsibilities, such as councils with councillors who remain active solely at a municipal and regional level. However, the Metropolitan Mayor can name a vice mayor, delegating specific issues to said vice mayor or to other metropolitan councillors. This happened in Milan, where the Metropolitan Mayor named a Vice Metropolitan Mayor and ten Councillors to handle the administrative aspects of the new government entity. The Metropolitan Cities are governed by two other collegial entities. The first is a Metropolitan Council, responsible for policy and control, which, depending on the population, is made up of 14–24 members who are not directly elected by citizens but rather "among and by mayors and councillors of the Municipalities included in the metropolitan area" (Vesperini 2014: 790). The second is a Metropolitan Conference made up of mayors from the Municipalities in the areas concerned, with proposal and advisory powers. In short, though the higher powers the mayors were given in the 1990s did not bring about significant improvements in the urban policies and more in general in the quality of life in large–medium-sized cities (Burroni et al. 2009; Piselli et al. 2012), Law 56/2014 still continues to consider this figure as the real *dominus* of the new Metropolitan Cities. Yet, the Metropolitan Mayor is apparently and paradoxically weak as it is the government structure of the Metropolitan City because of the poor democratic legitimacy (Onida 2015).

5.3.6 General Objectives

The Metropolitan Cities have three general objectives: strategic regional development; integrated promotion and management of services, infrastructure, and communication networks; and handling government relations with the other levels of local government and with other Metropolitan cities, also in Europe (section 1, paragraph 2). Compared to the general purpose of the Provinces, there seems to be an attempt to orient the Metropolitan Cities towards defining a new overall strategy for their territories that would overcome the sector-specific thinking as well as local management by Municipalities. At the same time, the effort is being made to put Metropolitan Cities into competition with other urban areas, a phenomenon that is becoming the norm throughout Europe and internationally. This competition—as we know—has to do with the capacity to attract investment for manufacturing, commercial, education, and research activities (Cremaschi et al. 2015). This seems to portend in Italy with two speeds: the one governed by the Provinces, and the other, by nature more dynamic and attractive, by the Metropolitan Cities. This is a situation that is not free from danger, and one that risks creating inequalities, at a national level, that are "antithetical to the goals of territorial cohesion introduced in the European Union with the 2007 Treaty of Lisbon and articulated in the Europe

2020 Report, both of which recognise the important role that an integrated, balanced, and inclusive territory plays" (Righini and Dallasta 2014: 3).

5.3.7 Special Functions

Beyond the general objectives, the law also more concretely establishes some of the fundamental functions that Metropolitan Cities have to perform. In particular, the "adoption and yearly updating of the three-year Strategic Plan of the metropolitan region, which sets the guidelines for the entity and for [...] the Municipalities"; "overall regional planning, including the communication, service, and infrastructure networks"; and, calibration of the "coordinated systems for the management of public services". They also have to deal with all matters concerning "transportation and roads"; the "promotion and coordination of economic and social development", and the "promotion and coordination of computerisation and digitisation in metropolitan areas" (section 1, paragraph 44). If we pause briefly to look at the aspects commonly considered as falling within the purview of urban planning, we notice at least two things. The first is the effort to overcome the distinction between the political administrative spheres and those specifically connected to regional planning. For example, the three-year Strategic Plan for the metropolitan area seems to be—as the Italian Law intends it and as it is interpreted at an international level (Perulli 2004)—a regional planning instrument that specifies political/economic/social objectives as well as objectives more directly related to urban design and regional planning. The second is the call for an overall regional plan distinct from the Strategic Plan, a choice that—one would assume—triggers a detailed urban planning exercise. However, key issues such as transportation and roads have been removed from consideration, as if they were irrelevant or foreign to the matter at hand.

5.4 Tools to Ensure Local Self-Government

5.4.1 The Regional Law

Unlike the similar French Law enforced in 2015, the pull in Law 56/2014 towards administrative decentralisation and local self-government does not provide for a real tax independence of the Metropolitan Cities (Gibelli 2015).[5] Independence drives apparently emerge in the potential offered by the Regions to define—through specific regional laws—the relations between the Regions themselves, the Metropolitan Cities, and the Municipalities. They also emerge in the opportunity

[5]On issues related to the Metropolitan Cities' funding, see also: Ferri (2008).

provided by the Metropolitan Cities to approve their own Statute suitable to regulate their operations.

Predictably, the first measure is giving rise to a range of situations at the national level. In Lombardy, the law that fully defines the competences and functioning of the Metropolitan City of Milan as well as its relationships with other institutions was approved in the autumn 2015. The Provisions for the development of the institutional role of the Metropolitan City of Milan (Regional Law 12 October 2015, n. 32) outline granting "the Metropolitan City of Milan additional functions beyond the fundamental ones" under Law 56/2014, including some functions previously attributed to the Province of Milan. On the other hand, other functions are unreasonably attributed again to the Region by this law. In fact, some sectors have remained completely excluded from the jurisdiction of the Metropolitan City such as agriculture, cultural policies, and environment and energy management, while others cannot aspire to the level of integration that Law 56/2014 seems to promote. This is the case with public transportation. The "Metropolitan City essentially takes over the responsibility for transportation outside the city limits much like the former Province did, yet it has little to say over transportation within the city limits or the rail lines [which interact with the city's public transportation network: these lines] […] remain firmly in the hands of the Regions" (Pizzul 2015: 9). In other terms, the Regional Law has not taken full advantage of the scale of the reform. Instead, it considers the Metropolitan City of Milan more or less in the same way as the Provinces, in other words, a body that coordinates with the Municipalities involved rather than a government entity able to act with real political autonomy. In the light of this, it appears that the "specificity of the Lombard context [has not been adequately interpreted. Indeed, perhaps this would have required the] creation of a Federation of Metropolitan Cities, in which the Regions [would have been able to be] recreated from scratch with the task of sorting and addressing issues in relation to the regional context, understood for what it is: a polycentric system of interdependent metropolitan entities" (Consonni 2014c: 10).

The direction of *Regione Lombardia* aimed to limit the pull towards administrative decentralisation in favour of Metropolitan City of Milan can also be noticed in other two aspects. The first concerns the creation of the Permanent Conference of the Region-Metropolitan City that was designed as a bridge between the two entities but that—according to some scholars—has actually created the basis for a 'mechanism of control' by the first over the second (Gibelli 2015). The second aspect concerns the decision to impose the fact that the Metropolitan Regional Plan is "drawn up on the basis of the principles and directions established by the Regional Plan". That is a decision that would apparently be logical, though on the one hand it denies the independence in the regional governance of the Metropolitan City while on the other hand it betrays the trends of the Lombardy urban planning. Over the last ten-year period, this has in fact travelled an opposite path, in its believing in urban planning by parts, out of a general planning framework and also notwithstanding the planning tools as approved at a local level (Brenna 2006).

5.4.2 *The Statute of Metropolitan City*

Law 56/2014—as we have said before—also calls for each Metropolitan City to have its own Statute. This instrument gives the local entity "unusual autonomy to better adapt to the specifics of its territory and resident community preferences" (Camagni 2014b: 3), regulating the organisation of the entities and defining the competencies of its bodies. The Statute of the Metropolitan City of Milan was drafted by the Metropolitan Council, supported by a committee of experts from various Milan Universities, and coordinated by the *Centro Studi per la Programmazione Intercomunale dell'area Metropolitana* (PIM) (Metropolitan Area Inter-Municipal Planning Research Centre) whose commitment in this field came before the coming into force of the law (Corvi and Sacchi 2013). It was approved by a large majority on 22 December 2014 by the Conference of Mayors of the Municipalities belonging to this catchment area. The definition and approval of the Statute of the Metropolitan City of Milan seem to be more a technical rather than a political reality. Its adoption has aroused a range of criticism, mostly related to the lack of participation and undemocratic nature of its creation. Some have noted that the Statute has been approved by administrative bodies that were not directly elected and therefore were not authorised to decide on such a significant deed for the community. Others have emphasised that the public has only been involved via an open call for input that very few have heard and that there has been no extensive public debate (see, for instance: Arrigoni 2014; Ballabio 2014; Meroni 2014).

In its seventy articles, the Statute of the Metropolitan City of Milan outlines its general principles. It also describes its bodies, functions, and objectives; and finally, it explains the organisation of the new administrative structure for the Milan territory. However, it is less precise about the method with which these decisions will be taken. The Metropolitan City of Milan has four overall objectives. At least one aspect of the first three has elements that directly relate to urban design and regional planning. These are as follows: "management and strategic development of the metropolitan territory, also through the integration of services, infrastructure, and communication networks" (SCMM, section 3, paragraph a); "development of the territory's manufacturing industry" (id., paragraph b); and finally, "sustainable development, overall urban quality, promotion of [...] the environment, and soil and hydrogeology protection, considered important resources that contribute to the unique identity of the metropolitan territories" (id., paragraph c). These strategic goals are achieved by integrating the policies and actions of heterogeneous spheres. The purpose is to give form to a regional government that, as opposed to what has happened so far in Italy, is not limited to sector-specific planning, but rather takes advantage of the potential generated by the ensuing interactions. One cannot help but notice the attention paid to relevant issues in current Italian and international urban planning: sustainable development, urban quality, the environment, and soil protection. These buzzwords have accompanied practically any plan or design in

Italy for years now, often without concrete actions being carried out. In fact, the state of the art in Milan and Lombardy on these same issues is clear indication that there is a very different reality in terms of urban and regional transportation systems, land use, quality of urban space, and the rural landscape.[6]

Another issue the Statute of the Metropolitan City of Milan places particular attention on is that of participation. This is not only in terms of urban design and regional planning, but more generally as a model of citizenship. For example, one section emphasises that “the Metropolitan City celebrates the right [of citizens] to participate in policy making and administration”, and not just those who are officially residents (section 3, paragraph c). In another, it stresses that “the Metropolitan City encourages citizens from individual as well as associated Municipalities to be directly involved in the choices made by the body” (id., section 7, paragraphs 1 and 2). Consequently, it should be possible to submit and file questions, applications, petitions, public proposals, and to announce various types of referendums (advisory, proactive, and repealing). Also called for is the creation of a Metropolitan Forum of Civil Society to encourage ongoing dialogue among different contingents within the Milan community (id., section 14). In essence—even if it seems premature to say how effective this instrument is and whether or not it is able to raise the level of democracy in the government and the regional planning process—we can state that the Statute of the Metropolitan City of Milan seems to have responded to and institutionalised the requests of those who pointed out the lack of citizen participation during the drafting process. We can also notice the attempt to correct, by means of this local tool, some faults in the national law.

5.5 Regional Government and Planning

5.5.1 The Strategic Plan

The first thing that the Metropolitan City of Milan must address—now that the Statute has been approved and its government bodies settled—is the definition of the Strategic Plan. Like the national law, the Statute considers this instrument to be the fundamental focus of all policies in the Metropolitan City. “It configures the scenarios and determines the overall development objectives of the metropolitan community and the associated socio-economic and regional-environmental sustainability” (section 34, paragraph 1). The Strategic Plan is considered a complex instrument designed to coordinate economic, social, environmental, and urban issues. It is supposed to contain “a long-term vision that defines a strategy to guide all the various aspects of regional development” (section 34, paragraph 2). It is also considered a superior instrument to all the others the Metropolitan City has to

[6]On the land use in Lombardy, see for instance: PIM-Atlante (2009), CRCS-Report (2014), ISPRA-Report (2014), ATPM, DUSAF, CRCS, ISPRA.

respect, with "the other planning acts and the other acts in general" having to substantially comply with it (section 35, paragraph 2). Yet, all these aspects seem to be in clear contradiction with the decision to attribute a 3-year length with yearly updating to the Strategic Plan.

Defining the Strategic Plan of the Metropolitan City of Milan—with an area of about 1600 square kilometres and more than three million inhabitants (ANCI)—means designing an innovative process and tools with little tangible experience to draw upon. If we think of the second post-war period, we could perhaps mention the experience of the *Istituto Lombardo Studi Economici e Sociali* (ILSES) or that of the *Piano Intercomunale Milanese* (PIM). More recently, between 2004 and 2009, the Province of Milan has tried to establish a strategic plan called *Città di Città* (Balducci et al. 2011). In other Italian Metropolitan cities, Turin and Bologna have already adopted a Strategic Plan. At a European level, we could consider the governance models of the metropolitan areas of Amsterdam, Barcelona, Glasgow, Lyon, London, Manchester, Vienna or Stockholm, whereas for the non-European cities we could consider the experience of cities such as New York, Detroit, or Rio de Janeiro, even if they are completely different realities compared to that of Milan. On the background, the recommendations emerged from international organisations such as the OECD and especially the EU, which, first with the 2007 Leipzig Charter, and then with the document Cities of Tomorrow (2011), and finally the Urban Agenda Europe Strategy 2020, provide a road map for the future of big cities and metropolitan territories (Pacente 2013; Vitali 2014).

On 16 April 2015, the Metropolitan Council of the Metropolitan City of Milan approved the Guidelines for the Preparation of a 2016–2018 Strategic Plan. In addition to its being understood both as a process and as a product, in principle the Plan should be participatory ("able to encourage inter-institutional and social protagonism through interaction and participation"), inclusive ("able to activate all the actors and energies in the metropolitan area"), innovative ("inspired and up-to-date with the best international practices"), concrete ("able to identify highly feasible and sustainable instruments, policies, projects, and actions"), selective ("able to prioritise projects, focusing on a few specific fields of action"), empowering ("able to remove development obstacles and create a working environment that facilitates action by public officials as well private and 'mixed' entities"), and finally ambitious ("able to act upon and reformulate the organisation of the metropolitan government structure as well as the technical-administrative apparatus according to goals and projects") (CMM-CM 2015).

On this basis, the Metropolitan Council decided to publish a Metropolitan Manifesto—intended as the "political cultural positioning document" of the Strategic Plan—and to define an agenda—intended as a way to choose priorities. Therefore, it has called for a phase of surveying the positions and local actors present on the territory, as well as a process of constructing the policies and techniques of the new instrument, before finally following a government approval process. To make this process as open as possible, there are a series of round-table discussions involving both government and non-government actors. This is followed by a Metropolitan Forum and various types of online participation. From the very outset of drafting the Plan, there is the opportunity for an extensive reworking of the hierarchy of typical urbanism elements which, but for this approach, would have been part of a strategic regional plan (transportation, housing, labour, and so on).

Early bottom-up reactions to the Strategic Plan taken from two documents (July and September 2015) by the *Centro Studi PIM* (or Metropolitan Area Inter-municipal Research Centre), which supports the Metropolitan City of Milan in defining the Plan, indicate the following:

(a) from the Municipalities:—First and foremost, the need for greater "clarity in the roles and functions of the Metropolitan City", an entity whose function continues to be perceived as hazy.—Second, with regard to the Plan, the need for an instrument with a "shared vision of medium to long-term regional development (at least 10 years)" and at the same time, "a high degree of activity in the short term". The Municipalities seem to be concerned about having lost a role in the planning process and about the restraints put on their activities.—Third, the issue of governance, i.e., the real opportunities for action by the Municipalities in the homogeneous zones and in their relations with the capital city, Milan, whose role appears to be felt as threatening and overbearing (CMM-PIM 2015).

(b) from the other socio-economic actors in the region who have been involved in this process (also through interviews): new approaches to the governance of complex issues; a Plan designed to increase the global competitiveness of the Metropolitan City of Milan; infrastructure, especially intangible ones. Generally, what the socio-economic actors hope for is a Metropolitan City rooted in innovation ("that focuses on young talent, creativity, and the university system"), attractiveness ("the capacity to attract talent and international capital"), competitiveness and a collaborative approach ("investment in the simplification and standardisation of regulations") (CMM-PIM 2015; see also: CMM 2015).

5.5.2 *The Metropolitan Regional Plan*

In addition to the Strategic Plan, the Metropolitan City of Milan is to equip itself with another instrument that is more explicitly related to urbanism: the Metropolitan Regional Plan. According to the Statute, it must adhere to the objectives of sustainable development and therefore be oriented towards the following: "strengthening and development of infrastructure and public transportation", "revitalisation of the built fabric", "strengthening and upgrading of public services and spaces", "construction of an ecological metropolitan network", "development and protection of the agricultural system and […] production, free soil, protected regional areas, metropolitan parks and landscape resources" (section 36, paragraph 2). After years in which Milan has headed in completely the opposite direction, finally, the Statute explicitly states that the Metropolitan Regional Plan considers "the soil to be a finite and irreplaceable resource" (section 36, paragraph 2). This is a controversial issue experiencing lively debate throughout Italy; it is also the subject of various proposed laws, including some under consideration in Parliament.[7]

The Metropolitan Regional Plan is not a simple guideline instrument for municipal plans in the metropolitan area. According to the Statute, it "produces direct effects on Municipalities and the private sector" (section 36, paragraph 4). As such, it will influence the actions of the Municipalities and the local community in diverse fields, including urban revitalisation policies, environmental protection, protection of landscape resources, and the programming of social housing construction. The issue of transportation, while covered by certain specific sections of the Statute (section 39 Road Networks and section 40 Public Transportation), is correctly redirected to the Statute itself in the Metropolitan Regional Plan. This instrument is also responsible for "metropolitan-level infrastructure programming, which also includes the communication, services network, and technological infrastructure of the metropolitan community" (section 36, paragraph 3). It is particularly interesting that the Statute considers alternative forms of transportation, creating incentives for so-called gentle and sustainable transportation (section 38, paragraph 4). Finally, the Metropolitan Regional Plan will have a direct effect on the economics of urban transformation, i.e. on the policies of "equalisation, compensation, and creating incentive on a regional scale for the purposes of pursuing a balanced distribution of advantages and disadvantages connected to the development and transformation of the region, also through inter-municipal taxation instruments" (section 36, paragraph 4).

It will therefore be an instrument that has simultaneously strategic, coordination, and prescriptive functions and above all, an instrument that from the point of view of contents is required to adopt "a regional strategy […] that is able to combine spatial visions and projects, namely real possibilities of transformation of physical space that are identified in investment potential and consistent functional programmes" (Pasqui 2015: 123). Unlike many plans from the second post-war period organised into an unlimited economic growth and building development perspectives, the Metropolitan Regional Plan is to consider both the long-term economic crisis that affects the Lombardy region—with clear effects on the physical context due to the increase in the unused craft, industrial, commercial, and residential buildings—and the crisis of urban and landscape spatiality that have a significant impact on the community life. From this viewpoint the need is strong for "pursuing urbanity [namely] the urban quality of places and relations (within an inclusive logic)" that is aimed to reduce inequality and promote social integration. At the same time "actions for the safeguard and treatment of landscape" cannot be put off any longer, in that they are oriented to "reduce entropy in the metropolitan

[7]See, for instance: Camagni et al. (2002), Bonomi and Abruzzese (2004), Gibelli and Salzano (2006), De Seta (2007), Tonon (2007), Bonomi (2008), Consonni (2008), Settis (2010), Giudice and Minucci (2011), Pileri and Granata (2012) and Romano (2013).

relations" (Consonni 2014a: 5).[8] The same is for actions aimed to fight urban sprawling, an irrational allocation of functions that is exclusively determined by market logics, the destruction of fertile soil.[9]

5.6 Conclusions

Whereas more thoughtful and structured assessments of recent governance and regional planning innovations in Italy will have to wait until the local government reorganisation process activated by Law 56/2014 is fully implemented and though it is clear that a single law is not enough to provide institutions with modern and effective governance and planning practices (in addition to innovative and well-devised instruments good policies are indispensable, and most of all an informed and engaged citizenry able to guide the local government in specific directions), we can conclude that this law is a reform with several limits from both the points of view of the regional and urban governance and planning. There are different reasons and the first probably concerns the fact that this law was specifically conceived to reduce public spending rather than to improve the regional government and design quality. The most critical issues regarding the regional design and urban and regional governance include:

- the superimposition of the regional boundaries of the old Provinces and those of the new Metropolitan cities: the objective should have been that of equipping Italy with new local government organs (and therefore new planning tools) in compliance with the current regional dynamics independently of the old administrative borders;
- the conflicting tendency to the union (of the Municipalities) and the partitions (of the chief towns) on the basis of merely administrative logics;
- the imbalance of powers between the Metropolitan cities and the Provinces that could increase the inequalities between regions that already are richer in terms of opportunities and resources and less lucky territories;
- the imbalance of powers between the chief town and the other Municipalities within the Metropolitan Cities areas that makes the regional governance democracy risky;
- the poor clearness of the scope of action of the planning tools provided for the Metropolitan Cities, even in connection with the existing tools that are still in force or those at a higher level;

[8]See also: Consonni (2008).

[9]On the Metropolitan City spatial planning theme, see also: Marson (2006).

- finally, the fact that the law—as many critics noted—is confused and contradictory in a sense; it involves particularly complex decision-making procedures that are difficult to be grasped by vast public, thus limiting the participation the law itself should pursue.

That said, and despite the limits outlined above (which could be better specified in a more extensive and detailed document), it appears clear that the opportunity offered by this reform should not be wasted, especially if one considers the impact it will have on the economy, manufacturing and culture of key Italian regions. It is hoped that the law is exploited to its full potential (especially the Statutes and the Strategic Plans), giving shape to effective and widespread governance and planning tools. An error in this regard would have serious consequences on the life of Italian citizens of today and tomorrow.

Acknowledgments My thanks to Elena Bertani, Sergio Brenna, Giancarlo Consonni and Anna Di Ciolo for discussing with me the content and the register of this chapter and for their suggestions.

References

Arrigoni F (2014) Città metropolitana. Cosa c'è dopo lo Statuto. ArcipelagoMilano 44(VI):5

Balducci A, Fedeli V, Pasqui G (2011) Strategic planning for contemporary urban regions: city of cities: a project for Milan. Ashgate, Farnham

Balducci A (2015) Le trasformazioni post-metropolitane e il modificarsi del legame tra spazio, forme dell'urbano e confini amministrativi. In: Lodigiani R (ed) Milano 2015. La Città metropolitana: sfide, contraddizioni, attese. Rapporto sulla città. Ambrosianeum Fondazione Culturale. Franco Angeli, Milano, pp 41–54

Bartaletti F (2009) Le aree metropolitane in Italia e nel mondo. Il quadro teorico e i riflessi territoriali. Bollati Boringhieri, Torino

Brenna S (2006) Milano "dall'esterno e da lungi". Gangemi, Roma

Ballabio V (2014) Città metropolitana o castello Kafkiano? ArcipelagoMilano 39(VI):9–10

Ballabio V (2015) Legge Delrio: va in scena il reality show "Città metropolitana". ArcipelagoMilano 36(VII):6–7

Beltrami Gadola L (2014) Milano città metropolitana: l'avvio di un pasticcio "retrò". ArcipelagoMilano 34(VI):2

Boeri S (2015) Interviewed by O. Liso. La Repubblica (ed. Milano) 30 January

Bonomi A, Abruzzese A (eds) (2004) La città infinita. Bruno Mondadori, Milano

Bonomi A (2008) Milano ai tempi delle moltitudini: vivere, lavorare, produrre nella città infinita. Bruno Mondadori, Milano

Burroni L, Piselli F, Ramella F, Triglia C (2009) Città metropolitane e politiche urbane. Firenze University Press, Firenze

Camagni R (2014a) Città metropolitane? No, solo province indebolite. Lavoce.info. http://www.lavoce.info. 18 Feb

Camagni R (2014b) Gli statuti metropolitani e l'esempio delle métropoles francesi. ArcipelagoMilano 35(VI):3–5

Camagni R, Gibelli MC, Rigamonti P (2002) I costi collettivi della città dispersa. Alinea, Firenze

Chiacchio T (2014), Disposizioni sulle città metropolitane, sulle provincie, sulle unioni e fusioni di comuni: L. 07 aprile 2014, n. 56. Commento a prima lettura. Giannini, Napoli

Consonni G (2008) La difficile arte: fare città nell'era della metropoli. Santarcangelo di Romagna, Maggioli
Consonni G (2014a) Città metropolitana. Idee forti e questioni istituzionali. ArcipelagoMilano 20 (VI):3–6
Consonni G (2014b) Città metropolitana e democrazia. Eddyburg. http://www.eddyburg.it. 17 July
Consonni G (2014c) Governo metropolitano: il necessario riequilibrio delle competenze. ArcipelagoMilano 33(VI):9–10
Consonni G (2014d) Sulla città metropolitana non si può più sbagliare. La Repubblica (ed. Milano). 8 Oct
Consonni G, Tonon G (2001) La terra degli ossimori: Caratteri del territorio e del paesaggio della Lombardia contemporanea. In: Bigazzi D, Meriggi M (eds) Storia d'Italia: Le regioni dall'Unità a oggi: La Lombardia. Einaudi, Torino, pp 51–187
Corvi D, Sacchi F (2013) Milano: prove di governo metropolitano. Urbanistica Informazioni 245–246(XXXX):32–33
Cremaschi M, Delpirou A, Rivière D, Salone C (eds) (2015) Métropoles et Régions: Entre concurrences et complémentarités: regards croisés France/Italie. Planum, Roma-Milano
De Seta C (2007) Bell'Italia: Patrimonio e paesaggi tra mali e rimedi. Electa, Milano
Ferri V (2008) Governare le città metropolitane: Un'istituzione del federalismo. Carocci, Roma
Gibelli MC, Salzano E (eds) (2006) No sprawl: Perché è necessario controllare la dispersione urbana e il consumo di suolo. Alinea, Firenze
Gibelli MC (2015) Grand Lyon Métropole e Città Metropolitana Milanese: Un confronto impari. Eddyburg. http://www.eddyburg.it. 25 Oct
Giudice M, Minucci F (eds) (2011) Il consumo di suolo in Italia: Analisi e proposte per un governo sostenibile del territorio. Esselibri, Napoli
Lanzani A (2004) Un'insolita rappresentazione per la governance della regione urbana lombardo-milanese. Territorio 29–30(II–III trim.):126–131
Marson A (ed) (2006) Il progetto di territorio nella città metropolitana. Alinea, Firenze
Mazza L (2015) Spazio e cittadinanza: Politica e governo del territorio. Donzelli, Roma
Meroni V (2014) Lo statuto della città metropolitana: Golia detta le regole. ArcipelagoMilano 39 (VI):10–11
Natale G (2014a) Di dubbio in dubbio: Ecco la città metropolitana. ArcipelagoMilano 34(VI):5–6
Natale G (2014b) La città metropolitana come sistema equilibrato dei Comuni. ArcipelagoMilano 38(VI):9–10
Onida V (2015) La Città metropolitana fra logiche istituzionali e "peso" della storia. In: Lodigiani R (ed) Milano 2015: La Città metropolitana: sfide, contraddizioni, attese. Rapporto sulla città. Ambrosianeum Fondazione Culturale. Franco Angeli, Milano, pp 33–39
Pacente C (ed) (2013) La città metropolitana nell'Unione europea: Programmazione comunitaria, esperienze europee a confronto e nuove prospettive. Egea, Milano
Pasqui G (2015) Ripensare le politiche territoriali: La sfida della Città metropolitana. In: Lodigiani R (ed) Milano 2015. La Città metropolitana: sfide, contraddizioni, attese. Rapporto sulla città. Ambrosianeum Fondazione Culturale. Franco Angeli, Milano, pp 119–135
Perulli G (ed) (2014) La città metropolitana. Giappichelli, Torino
Perulli P (2004) Piani strategici. Governare le città europee. F. Angeli, Milano
Pileri P, Granata E (2012) Amor Loci: Suolo, ambiente, cultura civile. Cortina, Milano
Piselli F, Ramella F, Burroni L (2012) Governare città: Beni collettivi e politiche metropolitane. Donzelli, Roma
Pizzul F (2015) Regione: Tarpare la Città metropolitana. ArcipelagoMilano 34(VII):9–10
Pompilio M (2014) Introductory paper presented to the conference: Governo e pianificazione territoriale di area vasta: Prospettive alla luce della L 56/2014. Fondazione Romagnosi, Pavia. 19 Nov
Pompilio M (2015) Legge Delrio: l'interpretazione intelligente. ArcipelagoMilano 8(VII):8–9
Rapicavoli C (2015) Il riordino delle province e la schizofrenia normativa. LeggiOggi. www.leggioggi.it. 12 Jan

Righini S, Dallasta P (2014) Chi pianifica la città metropolitana? ArcipelagoMilano 7(VI):2–3
Roma G (2014) Riempire i vuoti del governo territoriale. Urbanistica Informazioni 255 (XXXXI):7–9
Romano M (2013) Liberi di costruire. Bollati Boringhieri, Torino
Sarfatti (2014) Città metropolitana e municipalità: gestione dal basso. ArcipelagoMilano 43(VI):7–8
Settis S (2010) Paesaggio Costituzione cemento: La battaglia per l'ambiente contro il degrado civile. Einaudi, Torino
Soldavini P (2014) Rigenerazione urbana strategica. Il Sole 24 ore (Nòva24 Tech). 17 Dec
Tonon G (2007) Il paesaggio umiliato: Insostenibile bruttezza delle metropoli. Ogni uomo è tutti gli uomini, Bologna
Tortorella W, Allulli M (2014) Città metropolitane: La lunga attesa. Marsilio, Venezia
Vesperini G (2014) Il disegno del nuovo governo locale: Le città metropolitane e le province. Giornale di diritto amministrativo 8–9(XXI):786–794
Vitale M (2015) Interview by O. Liso. La Repubblica (ed. Milano). 16 Jan
Vitali W (ed) (2014) Un'Agenda per le città: Nuove visioni per lo sviluppo urbano. Il Mulino, Bologna

Reports and Documents

CM-PM (2013) Comune di Milano, Politecnico di Milano—DASTU, Verso le nuove municipalità. Un atlante. Research coordinated by G. Pasqui e V. Fedeli, Milano
CMM (2015) Città Metropolitana di Milano. Milano metropoli reale, metropoli possibile: mappa delle idee. Verso il Piano Strategico Milanese (approved in September)
CMM-CM (2015a) Città Metropolitana di Milano, Consiglio Metropolitano. Approvazione linee di indirizzo per la predisposizione del Piano Strategico 2016–2018 […], session of 16 April 2015
CMM-CM (2015b) Città Metropolitana di Milano, Consiglio Metropolitano. Approvazione della proposta di costituzione e delimitazione di zone omogenee della Città metropolitana di Milano […], session of 17 September 2015
CMM-CM (2015c) Città Metropolitana di Milano, Consiglio Metropolitano. Approvazione della proposta di Regolamento per il funzionamento delle zone omogenee […], session of 17 settembre 2015
CMM-PIM (2015a) Città Metropolitana di Milano, Centro Studi PIM. Verso il Piano Strategico Metropolitano milanese: tavoli territoriali: prime indicazioni dai Comuni (approved in July)
CMM-PIM (2015b) Città Metropolitana di Milano, Centro Studi PIM. Verso il Piano Strategico Metropolitano milanese: tavolo metropolitano e interviste: prime indicazioni dagli attori (approved in September)
ISPRA-Report (2014) Istituto Superiore per la Protezione e la Ricerca Ambientale: X rapporto, 2014. Ispra, Roma
CRCS-Report (2014) Centro di Ricerca sui Consumi di Suolo: rapporto 2014. Inu edizioni, Roma
PIM-Atlante (2009) Centro Studi per la Programmazione Intercomunale dell'area Metropolitana (2009), Atlante del consumo di suolo in provincia di Milano. Series: Quaderni del Piano territoriale, n. 28. Provincia di Milano, Milano
SCMM: Statuto della Città Metropolitana di Milano (approved by the Metropolitan Conference of Mayors dated 22 December 2014 with resolution n. 2/2014)

Web Sites and Database

ANCI (Associazione Nazionale dei Comuni Italiani). http://www.anci.it. Accessed 15 Oct 2015
ATPM (Atlante dei Territori Post-Metropolitani). http://www.postmetropoli.it/atlante. Accessed 10 Oct 2015

CM (Comune di Milano). http://www.comune.milano.it. Accessed 20 Oct 2015
CMM (Città Metropolitana di Milano). http://www.cittametropolitana.mi.it. Accessed Sept–Oct 2015
CRCS (Centro di Ricerca sui Consumi di Suolo). http://consumosuolo.org. Accessed 15 Sept 2015
DUSAF (database on Destinazione d'Uso dei Suoli Agricoli e Forestali della Regione Lombardia). http://www.pim.mi.it/sit/dusaf/dusaf.html. Accessed 20 Oct 2015
GI (Governo Italiano. Presidenza del Consiglio dei Ministri). http://www.governo.it. Accessed 15 Oct 2015
ISPRA (Istituto Superiore per la Protezione e la Ricerca Ambientale). http://www.isprambiente.gov.it/it. Accessed 10 Sept 2015
PIM (Centro Studi per la Programmazione Intercomunale dell'area Metropolitana). http://www.pim.mi.it. Accessed Sept–Oct 2015
RL (Regione Lombardia). http://www.regione.lombardia.it. Accessed Sept–Oct 2015
Treccani (Istituto della Enciclopedia Italiana fondato da Giovanni Treccani), Vocabolario. http://www.treccani.it. Accessed 2 Sept 2015
Treccani (Istituto della Enciclopedia Italiana fondato da Giovanni Treccani), Enciclopedia. http://www.treccani.it. Accessed 2 Sept 2015

Part II
Eu Policies, The Urban Agenda and Local Governance

Chapter 6
Urban Dimension of the European Policies and the New EU Urban Agenda in Croatia

Dana Dobrić

Abstract The role of sub-national levels of government has significantly changed under the influence of Europeanization and so too the institutional frameworks of urban governance. Local units are expanding their activities to the European political arena, and their influence in strengthening local development is growing. Also, public management reforms are emphasizing the economic role of local government units and their active participation in stimulation of local development. The chapter addresses two main research questions: (1) What are the EU and Council of Europe recommendations for good local governance? and (2) Is Croatia implementing the European trends in urban governance? The main premise is that the process of European integration and institutionalization of relations between the European institutions and the Republic of Croatia has initialized the implementation of structural and functional reforms in the local government. The main hypothesis is set affirmatively: It is assumed that the majority of recommendations for good local governance have been adopted in the Croatian legal framework. For the purpose of hypothesis testing the legal analysis, teleological and linguistic method has been used on European and national legal documents (hard and "soft" law). The research has shown that: (1) There are 16 European principles for good local governance, and (2) all of them are implemented in Croatian legislation through a number of different standards. Therefore, the hypothesis has been proven. The research confirmed that the formal requirements for urban governance in Croatia are provided. In addition, the chapter recommends empirical research of practical implementation of the principles in everyday activities of the local institutions.

Keywords Local government · Urban governance · Development policies · Strategic planning · Participatory democracy · Croatia

D. Dobrić (✉)
University of Rijeka, Rijeka, Croatia
e-mail: dana.v.dobric@gmail.com

© Springer International Publishing Switzerland 2017

C. Nunes Silva and J. Buček (eds.), *Local Government and Urban Governance in Europe*, The Urban Book Series, DOI 10.1007/978-3-319-43979-2_6

6.1 Introduction

At the end of the 1980s, the intensification of the European integration process began, which ultimately led to significant changes in the status and role of the sub-national levels of government and through that on urban governance processes. By implementing the European legislative framework, local government units, particularly the cities, have initiated a number of structural and functional reforms in order to achieve stated goals such as effectiveness, responsiveness, and quality of service delivery. In addition, the new public management reforms emphasize the economic role of sub-national levels of government and the necessity to direct their activities toward the promotion and/or strengthening of local development and growth. Although the European regional policy is primarily geared toward strengthening the position of the regions, the importance of the local levels of government cannot be ignored. Cities, which represent larger self-governing units, are economically more prosperous and are therefore progressively taking the lead in the economic development of a wider territory. In this regard, European institutions recommend strategic planning and participatory democracy as the instruments necessary for achieving economic prosperity, social cohesion, and environmental sustainability (see: Allen 2000; Kearney 2013). The main purpose of the application of these instruments is to achieve a greater effectiveness in the process of political decision-making and to raise the awareness of the diversity of problems of local communities.

The Republic of Croatia started a political dialogue with the Council of Europe (CoE) in 1991, and the main goal of this cooperation was a prerequisite for a full membership in the council. The goal was achieved in 1996 after which a long process began of adaptation of national legislation to the obligations which Croatia undertook by signing hundreds of international treaties and partial agreements, especially in the field of local governance. Also, during the last 10 years, a process of harmonization of the national legal framework has been conducted and the implementation of measures for the adoption of *acquis communautaire* of the European Union (EU).

The central interest of this chapter is to establish the degree of Europeanization of the national system, i.e., the compliance of the European and national regulation of urban local governance and development policies. Therefore, the principles and standards proclaimed in the documents of the CoE and EU have occurred as a preliminary research matter. The main hypothesis is set affirmatively. That is, it is assumed that the majority of recommendations for good local governance have been adopted in the Croatian legal framework.

The testing of this hypothesis was carried out in two phases, primarily by using the legal–analytical method and the linguistic and teleological method with the aim of interpreting legal texts. In the first phase, legal documents of the CoE and EU were analyzed which in their title or keywords relate to some aspect of urban management. The expected result of the first phase is a catalog of principles and standards applicable to the design of urban policies.[1]

Compared to the EU, a far larger amount of documents of the CoE was analyzed, thereby taking into account its active role in regulating local governance. The attention was largely devoted to international agreements. Also important is the action of the Congress of Local and Regional Authorities of the CoE since the main activity of its committees is the adoption of resolutions and recommendations for Member States and the sub-national levels of government.

On the other hand, the EU has exclusive competence for regulating the system of the Member States' local self-governments. Thus, the primary and secondary EU legislation does not regulate this area systematically but through certain provisions. Such provisions are found in the acts governing other areas and policies, in the framework of those issues that may be important for sub-national governments. Likewise, EU institutions have adopted thousands of soft law instruments, such as communications, resolutions, and opinions that regulate various topics from the sphere of local self-government but do not produce legal but practical effects.[2]

The first section of the chapter analyzes the European (European Union and Council of Europe) policies and standards for good local governance, with special emphasis on local development policies and the role of European cities in their implementation. The implementation of the European *urban acquis* into the legal framework of the Republic of Croatia is explored in the second part where the focus is placed on the measures which Croatian cities have adopted so as to encourage local development and growth. The concluding part of the chapter contains comments gained from insight and a critical review of the involvement of Croatian local

---

[1]Legal norms can be classified into three categories: rules, standards, and principles. Rules and standards are the framework that is sufficient to address a specific legal dispute. Principles, in turn, form socially acceptable values and practices; their purpose is not to resolve legal issues, but they represent guidelines for the interpretation or application of rules and standards (Alexander 2013; Alexy and Rivers 2010; Dworkin 2003; Schlag 1985; Shapiro 2007).

[2]Senden (2004, 2005) has developed a concept, classification, and action of the so-called soft law of the European Union. By identifying three key elements of non-binding instruments (they refer to the rules of procedure or assumption of certain liabilities; they produce certain legal effects and can have practical effects or impacts on activities), she has offered a definition of soft law: "rules of procedure contained in instruments that are not characterized as legally binding, but have certain indirect legal effects and are aimed at creating a certain practical effects" (Senden 2005: 22). Preparatory and informational instruments such as the green and white papers of the European Commission, action programs, and informative communication are adopted in order to prepare for a future regulation, or policy, or providing information about a particular action of the European Union, which gives them a "pre-legal function" (Senden 2005: 23). Communication[2] and information of the European Commission, guidelines, codes, and frameworks are categorized as interpretive and decisive instruments. Within the framework of their "post-law function" (Senden 2005: 24), their purpose is to interpret the norms of the legal system of the European Union and to guide their accurate application. Finally, steering instruments are focused on setting goals or giving effect to European policies. In doing so, formal acts are divided into recommendations (provided in Art. 288 TFEU) and informal acts into declarations, conclusions, resolutions, and codes. Given that they are used in order to harmonize different systems of the Member States, they have a "para-law function" (Senden 2005: 24).

units in the implementation of European trends. Also, proposals are given for the strengthening of the economic role of sub-national levels of government and a further adaptation of Croatian cities to the European development policy.

6.2 Council of Europe *Urban Acquis*

The Congress of Local and Regional Authorities adopted in 1992 the European Urban Charter (Charter I) as a result of previous activities of the council related to the design of urban policies, among which it is necessary to point out the program which was in the period from 1982 to 1986 developed by the Intergovernmental Committee composed of representatives of national ministries responsible for urban issues. Since 1986, the program has been delegated to the jurisdiction of the Congress in accordance with the trend of decentralization of decision-making on public policies in the Member States, which has resulted in an increase in the powers of local authorities in the field of urban issues.

Charter I is a manual on city management for local authorities and a practical instrument that forms the basis for the development of an international system of rewarding cities that accept the application of the principles of the Charter. The selection of principles listed in Charter I suggests a strongly pronounced dimension of local authorities since precisely this level is responsible for the development of cities. The attention is thereby paid to the qualitative aspects of urban development and the quality of life in cities, with special emphasis on the protection of fundamental freedoms and human rights.

By strengthening the process of globalization, cities are becoming the key actors in Europe's adaptation to new technologies and the environmental, economic, and social circumstances. In response to these changes, in 2008, the Congress adopted Recommendation n. 251 on the European Urban Charter II—"Manifesto for a new urbanity" (Charter II), which updated the previous, Charter I. As principles and standards that the Charters proclaim, the following can be singled out: (1) principle of non-discrimination designed through standards of antidiscrimination legislation and multiculturalism; (2) cohesion, through the standards of housing policy, intergenerational solidarity, and social sensitivity; (3) sustainability, through environmental standards, energy efficiency, alternative means of transport, and the use of green technologies; (4) participation, through the standards of founding citizen councils to participate in local elections, local referendums, and other forms of direct participation of citizens, civil monitoring, and evaluation of the work of local authorities; (5) partnership, through cooperation standards among local governments, public and private sector, cooperation between the local, regional, and national, and European level on the basis of a balanced partnership; (6) subsidiarity, through the standards of functional and fiscal decentralization; (7) transparency, through the standards of public consultation on urban projects and information availability to citizens on local policies; and (8) effectiveness, through the standards of the citizens' involvement in public affairs and the new information and communication technology.

Furthermore, the Center of Expertise for Local Government Reform[3] adopted in 2007 the Strategy for Innovation and Good Governance at Local Level. The Strategy is based on the lessons and experiences of the Member States, their cooperation within the council, and it contains twelve principles of good democratic governance, which constitute a guide for public bodies to improve the quality of local governance: (1) fair conduct of the election, representation, and participation, (2) responsiveness, (3) efficiency and effectiveness, (4) openness and transparency, (5) the rule of law, (6) ethical conduct, (7) competence and capacity, (8) innovation and openness to change, (9) sustainability and long-term orientation, (10) sound financial management, (11) human rights, cultural diversity, and social cohesion, and (12) accountability. Given the fact that the local level is closest to its citizens and provides them with basic public services, it is extremely important to adopt the principles of good local governance. Therefore, national and regional authorities and associations of local authorities are asked, in accordance with their responsibilities and powers, to undertake their activities toward the achievement of democratic governance. Development of an action program for this purpose is an expression of public commitment to improving the quality of governance and public services.[4]

Finally, the bodies of the council have adopted a number of recommendations, resolutions, and opinions in the field of urban policy and urban management. However, due to size restrictions, this chapter will not separately analyze each of these documents. Below are some of the most important recommendations.

Democratically elected local authorities are in the best position to address the problems of the urban society, the implementation process of planning, and control development. Within the framework of urban planning, they should take into account the insurance of a pleasant urban environment and neighborhoods, improvement of the environmental protection, maintenance and creation of open and public spaces, social protection, economic development, employment, adequate housing, a well-organized public transport, protection of monuments and cultural heritage, education opportunities, and the recreational and sports infrastructure. In doing so, the legitimacy of political decision-making of city authorities is maintained by public involvement in the management and for the participation of

[3]The center is the executive body of the council in the area of multilevel governance that supports central, regional, and local authorities to improve their legislation, institutional capacity and efficiency in service delivery, and the achievement of public authority. Also, through its action, it seeks to underline to the Member States the importance of the process of decentralization and strengthening of the capacity and role of local authorities. For this purpose, it offers, in accordance with the good practice of the EU, a series of practical programs and instruments in different areas: performance management, citizen participation, education and training, and exchange of experience. 32 programs have currently been implemented in twenty European countries. More information available at: http://www.coe.int/t/congress/Activities/partnerships/default_en.asp?mytabsmenu=3.

[4]European Committee on Local and Regional Democracy (2007).

citizens to be successful; they must be well acquainted with the ways of city management. Participation must be real, which means it is not enough to include citizens in the work of an advisory body with no real possibility of influence in the strategic decision-making.[5]

Following the adoption of decisions, strategies, and action plans for their achievement, a number of factors can contribute to a more successful implementation of the planned measures. The first position is funding. City authorities should ensure greater opportunities for revenue collection within their own local unit so as to be less dependent on central government funding. Funds which they collect should be sufficient to carry out the powers of self-government and transferred competencies. However, national authorities should regulate the legal framework for the allocation of the necessary financial support to city authorities for the purpose of building and maintaining the infrastructure of national importance.[6] Furthermore, city authorities should in the best possible way take advantage of the new information and communications technology in creating an efficient and transparent administration which will provide quality services. Administrative bodies are recommended to use new managerial methods of action with a warning that their implementation must not be at the expense of the democratic and public control of community governance and of which there is a risk especially in large urban agglomerations.[7] The last factor is the relationship which the city government maintains with other stakeholders. For example, it would be useful to enter into partnerships with the business and voluntary sector, to maintain permanent contacts with other cities in order to share experiences and best practices in management and to build international relations and cooperation. Also, with the aim of promoting the economic development and reconstruction of their communities, local governments should support the establishment of an agency that would collaborate with the market and industry sector.

6.3 European Union Urban Acquis

Within the European Union, standards of urban policies, in addition to institutions, have also been shaped by the Ministers of the Member States in documents adopted at informal meetings, such as the Lille Action Program (2000), the Rotterdam "*Urban Acquis*" (2004), the Bristol agreement (2005), and the Charter of Leipzig (2007). Good local governance, strengthening of their economic role, and encouragement of an integrated urban development are proclaimed with the above-mentioned documents.

[5]The Congress (1995).

[6]The Congress (2003).

[7]The Congress (1996).

The principles of good local governance[8] are built on the recommendations contained in Charter I and II, and it is emphasized that their use contributes to achieving eight goals of European urban policy[9]: (1) better understanding of the role of cities in urban development, (2) development of an integrated, results-oriented urban strategy, (3) ensuring the participation of local actors, (4) prevention of segregation, (5) integrated balanced urban development, (6) partnership between the private and public sector, (7) strengthening of European cooperation, and (8) use of modern information technology in the formation of urban policy. In doing so, cities, confirming their economic position, are given the importance of knowledge centers, innovation sources, and economic growth.

Due to the existence of substantial differences in terms of economic and social opportunities, and the environmental quality in some urban areas, a number of documents proclaim integrated urban development. The policy of integrated urban development involves a fair consideration of all the issues and concerns that are relevant for urban development. It is a process of coordination of spatial, sectoral, and temporal aspects of the different areas of urban policy in which the participation of economic and other stakeholders, as well as the general public, is of utmost importance. For cities, which still have not developed such a practice, it is advisable to develop a program of integrated urban development and to establish a mutual cooperation and equal partnership with the regions and rural areas. More closely connected cities at the European level would contribute to the realization of an economic, social, and territorial cohesion.

From the recommendations, resolutions, and opinions of the institutions of the Union, it may be concluded that public and private actors at the local level have a key role in the effective implementation of development policies and programs. Depending on the circumstances in each Member State, connecting the public sector and development process may result in the need to increase the capacity of local authorities and in the strengthening of local governance structures.[10]

---

[8]The white paper of the European Commission on European Governance contains the principles of good governance applicable at all levels of government (global, European, national, regional, and local). According to the principle of responsibility, each institution of the Union, Member State, and other actors involved in the development and implementation of European policies is responsible for their actions, i.e., failures to act. The principle of participation implies the involvement of all stakeholders in the decision-making process, while the principle of openness implies providing active communication between the institutions of the Union and the Member States about the activities and decisions of the Union (mechanisms of consultation and dialogue should be accessible and understandable to the public). The principle of the effectiveness of management involves the application of effective and timely policy measures, and the latter principle applies to cross-compliance, easy understanding of policy measures, and actions (European Commission 2001).

[9]*Urban Acquis* (2004).

[10]Limited capacity issues should be solved at the individual, institutional, and organizational level. At the individual level, local capacity may be limited by a lack of skills, qualifications, or, in turn, certain attitudes of local officials and civil servants. In institutional terms, the capacity depends on the structures and procedures that define and govern staff activities in local authorities. Regarding the organizational level, the capacities may be affected by the legal framework, policies, and rules

Good governance at the local level is a prerequisite for achieving sustainable development, i.e., creating conditions for inclusive, responsive, and efficient development processes.[11] On the one hand, the quality of local governance depends on the political will of the central government through regulatory instruments that create an enabling environment at the local level, in particular by ensuring a sufficient degree of autonomy of local authorities regarding the implementation of their powers.[12] Also, the quality of governance depends on the manner of implementation of public policies, the provision of public services, and the interaction with other public institutions, citizens, and the private sector and on the method of allocation of available resources.

The European Commission (hereafter: the Commission) argues for the application of the territorial approach to development. The main feature of this approach is a long-term and dynamic bottom-up process in which different sectors and groups of stakeholders (i.e., multisectoral approach with multiple stakeholders) jointly define development priorities and shape and implement development strategies.[13] The Commission points out that such an approach can contribute to improving the quality of local life,[14] to achieving a balance between the socioeconomic growth, equity, and environmental quality, as well as to an increase in the resilience of the most vulnerable social groups. The territorial approach can also affect the local political process and increase transparency and continuity of monitoring development processes toward sustainability.[15]

(Footnote 10 continued)

governing the attitude of local authorities toward other actors (Communication from the Commission 2013). The European Union should, in addition to national governments, also invest in developing the capacity of local authorities and their cooperation with other local actors. European cities are considered to be the drivers of economic growth and development because they have the knowledge and capabilities that are needed to establish competitive local, regional, and European economies (Committee of Regions 2002). Consequently, URBACT program for exchange and learning between European cities which promote their sustainable development is financed from the European Regional Development Fund. This confirms the important role that cities play in dealing with the complex contemporary social change. Efforts are being made to encourage them to find new and sustainable solutions that integrate economic, social, and environmental dimensions. More information available at: http://www.urbact.eu/en/homepage-2.

[11]European Parliament (2007).

[12]Many Member States, in terms of development processes, have granted certain powers and responsibilities to the sub-national levels of government (territorial approach to policy development). However, political recognition is not always followed by an appropriate autonomy level, development capacities, and financial resources. Therefore, it is necessary to overcome a number of obstacles in order to achieve a development potential of local authorities.

[13]Local units have a strategic role of a mediator between national authorities and civil society organizations, as well as promoters of consultation between various partners in development processes (European Commission 2010b).

[14]As territorial levels closest to the citizens, information on the needs and interests of local people is available to them to the greatest extent.

[15]Communication from the Commission (2008, 2013).

Sustainable economic growth and a knowledge-based economy should contribute to the growth rates of employment and social cohesion (Bache 2007). In order to achieve this, employment policy should be shaped in accordance with innovation and integration policies, entrepreneurship, information society, and new economic sectors. For the purpose of effective implementation, it is necessary to ensure adequate coordination of these interrelated policy areas and the involvement of various local actors: companies, educational institutions, social partners, private and public employers, NGOs, and the public. It is a partnership dimension, i.e., a participatory approach to management development, according to which different groups of partners work in a complementary and coherent manner.[16]

The quality and effectiveness of the partnership will be higher if they are concluded for a longer period of time and if they demonstrate a high level of flexibility. Taking into consideration different socioeconomic conditions of various levels of government, a flexible approach is an important precondition for a successful implementation of territorial partnerships. This particularly applies to the allocation and management of local and regional resources intended for a development cooperation.[17]

European public policies, including local policies, are formed in accordance with the Strategy for a smart, sustainable, and inclusive growth ("Europe 2020"; hereinafter: the Strategy).[18] Its main goals are to build an economy based on knowledge and innovation, effectively use competitive and "green" resources in all sectors, and to achieve high levels of employment that contribute to the social and territorial

[16]European Parliament (2007).

[17]European Parliament (2010) and Committee of Regions (2002).

[18]The Strategy builds on the Lisbon agenda (hereinafter: Agenda) and introduces new guidelines for strengthening the competitiveness of the European economy which has been affected by the current financial crisis. The Agenda has been adopted by the European Council as a strategic document containing the development objectives of the Union for a period of 10 years and a plan for their achievement. Strengthening the competitiveness of the knowledge-based economy, sustainable economic growth, and increase of employment rates and economic and social cohesion are attempted to be achieved through the: (1) development of the information society, (2) creation of a European space for research and innovation, (3) launch and development of innovative entrepreneurship (especially small and medium enterprises), (4) completion of the construction of the internal market of the EU, (5) effective and integrated financial markets, (6) coordination of macroeconomic policy, and (7) development of effective employment policies, improving the social protection system and promoting social inclusion. Because of the unsatisfactory results of its implementation in 5 years, particularly with regard to employment, the Agenda was revised in 2005 and three new objectives have been determined: to make Europe a more attractive place for investments and work, to provide knowledge and innovation for growth, and to devise policies that will target the creation of new and better jobs (Communication from the Commission to the Spring European Council 2005).

cohesion.[19] Ensuring their realization is one of the priority tasks of the Committee of the Regions (hereafter: CR), which provides supports and gives a policy contribution to the implementation of the Strategy.[20]

6.4 Results of the Research Phase 1

After the first phase of the research, it was possible to develop the concept of integrated urban development, which covers a number of aspects of good local governance, such as economic development, social inclusion, sustainable development, exchange of experiences and best practices, networking, participation, smart growth and development, urban planning, and so on.

Furthermore, it was found that a possible top-down Europeanization of the local level can be monitored in three dimensions of domestic structures according to which we have classified the identified principles (and standards): polity, policy, and politics dimensions.[21] In polity, or the dimension of political structures,

[19]Firstly, smart growth means strengthening knowledge and innovation; it requires improving the quality of education, strengthening the impact of research, promoting innovation and knowledge transfer in the EU, a better use of information and communication technologies, and ensuring that innovative ideas are transformed into new products and services that result in growth and quality jobs, as well as help in dealing with the European and global social challenges. Secondly, sustainable growth includes the construction of a sustainable and competitive economy which effectively uses its resources, the development of new processes and technologies, including "green" technologies, the acceleration of the development of smart grids by using information and communication technologies, the use of networks at the Union level and strengthening the competitive advantages of doing business, primarily in production and in small and medium enterprises, as well as helping consumers to cost effectively use the resources. Finally, inclusive growth means high levels of employment, investing in skills, fighting poverty, and modernizing labor markets and training systems, and the social protection in order to provide assistance to persons in predicting and managing change and building a connected society (Communication from the Commission 2010a).

[20]Important instruments for the achievement of the Strategy objectives by local and regional authorities are the European funding programs. Therefore, the Committee began its analysis of the proposed new policy of financing by giving opinions to draft the relevant European legislation. Cohesion policy is designated as a key investment instrument in the negotiations on the EU budget in which the Committee stresses the importance of a partnership approach. The measures envisaged to achieve the objectives of the Strategy are as follows: (1) monitoring the impact of the policies proposed in the Strategy, (2) proposing innovative ways of their implementation, (3) information and mobilization of local and regional authorities to achieve the objectives, primarily the exchange of best practices, (4) supporting the successful implementation of the internal market, and (5) conducting research ways in which culture can be a driver of growth and employment (Committee of Regions 2013).

[21]Dimensions of the Europeanization of national systems are the discussion topic of numerous studies. Domestic political institutions, intergovernmental relations, the judicial system, public administration, economic institutions, the relationship between government and society, state traditions, and collective identities are understood under the term polity, or the structural dimension. Policy, or the functional dimension, includes standards, instruments, approaches to

European institutions proclaim the principles of the local self-government, subsidiarity, partnership/mutual aid, structural adjustments, and participation. With regard to policy, or the dimension of public policy, the following principles were identified: sustainability, competitiveness, antidiscrimination, partnership, exchange of best practice, transparency, and effectiveness. In politics, or the dimension of political processes, the principles of direct and representative democracy are highlighted. The identified principles of *urban acquis* are described in Table 6.1.

The specific application of the principles is achieved by implementing a set of standards: For example, (1) the principle of local self-government can be implemented through the standards of decentralization, judicial protection of rights of local self-government, the special status of the major cities, and the development of strategic planning capacity of local government units; (2) the principle of non-discrimination through the guarantee of equal rights of all citizens in the local self-government and multicultural integration; (3) the principle of participation through intermunicipal cooperation, consultation and dialogue between local authorities, networks and associations of local authorities and local and national authorities, advisory bodies of foreigners and youth councils at the local level, and citizens' control over the actions of local authorities and administrative bodies; (4) the principle of partnership through twinning, cooperation of local authorities with the private sector and nongovernmental organizations; (5) the exchange of best practices also through twinning and membership in national and international associations of local authorities; (6) the principle of subsidiarity through a functional and fiscal decentralization and providing information to citizens on the activities of local authorities; (7) the principle of transparency guarantees access to information and official documents and consultations and dialogue with the citizens; (8) the principle of effectiveness also through consultations and dialogue with the citizens, entrepreneurial sector, nongovernmental organizations, and e-governance; and (9) the principle of sustainability through strategic planning development, an efficient use of energy resources, and environmental protection (Dobrić 2014).

6.5 Croatian Legal Framework on Urban Management

Based on the pre-established catalog of European principles and standards of urban management, their implementation into the national legislation was analyzed. It was attempted to find the Croatian equivalent to European recommendations in the

(Footnote 21 continued)

problem solving and public policy discourses. Finally, the politics dimension is oriented toward activities of political parties and interest groups, i.e., processes of forming, aggregation, and representation of the interests and public hearings. (Börze and Risse 2003; Kassim 2005; Katenhausen and Lamping 2003; Ladrech 2002, 2005, 2009, 2010; Mair 2007; McCormick 2011; Sošić 2006).

Table 6.1 Good local government principles encompassed in the EU and the CoE instruments
Source Dobrić (2014)

Principle	Description	EU	CoE
Equal access	Organization of public services at the level closest to citizens in order to meet the request for ensuring equal rights to public services	1	1
Quality of service	Compliance with standards and procedures that regulate public services and achieving maximum satisfaction of users	1	1
Competition	Guarantee of free movement of services and establishment in all Member States	1	0
Local self-government	The right and the ability of local authorities, within the limits of the law, to regulate and manage a substantial share of public affairs under their own responsibility and in the interests of the local population	0	1
Antidiscrimination	Equal opportunities for realization of human rights and fundamental freedoms without discrimination on any grounds	1	1
Accountability	Local government units are responsible for violations of constitutional and legal rights of citizens and the violation of the rule of law within their functioning	0	1
Sustainability	The balance between economic, social, and environmental requirements based on the effective management of natural and energy resources and green and competitive economy	1	1
Participation	Citizen participation in local public affairs as well as participation of local government units in European policy processes	1	1
Partnership	Vertical and horizontal cooperation to achieve objectives that are difficult or impossible to achieve without such cooperation	1	1
Legal certainty	Compliance with standards and procedures that regulate public services and stabilization of expectation of users	1	0
Exchange of best practices	Cooperation of local government units at the national and/or European level in order to find the best solutions in the application of European policies	1	1
Subsidiarity	Public responsibilities shall generally be exercised, in preference, by those authorities that are closest to the citizen, taking into consideration the extent and nature of the task and requirements of efficiency and economy	1	1
Transparency	Information about local public affairs and official documents of local authorities should be made public	1	1
Effectiveness	Implementing EU policies in a proportionate manner and taking decisions at the most appropriate level	1	1
Inclusiveness	Promoting full employment economy with the aim of achieving social and territorial cohesion and reduce the risk of exclusion of socially disadvantaged groups	1	1

(continued)

Table 6.1 (continued)

Principle	Description	EU	CoE
Mutual aid	Cooperation of local government units at the national and/or European level for conducting the affairs of common interest	1	1

Indicator "yes" = variable 1
Indicator "no" = variable 0

Constitution of the Republic of Croatia, constitutional and ordinary laws, regulations, strategies, agreements, and programs. The results are very extensive, and only the main conclusions will be summarized for the purpose of this chapter.

Firstly, regulations in which the principles of *urban acquis* are implemented in most of the provisions regulate the system of local and regional self-government (2001), regional development (2014), the special status of major cities (2001), local elections (2012) and the referendum (1996), budget (2008), financing of local government units (1993), free access to information (2013), the environment (2013), public–private partnership (2012), and the use of structural instruments of the EU (2012).

Secondly, in the context of the economic role of local government units and local development policies, the most important is the Regional Development Act. The Croatian regional development policy includes urban development. Its main goal is the achievement of economic growth and development of the country in accordance with the principles of sustainable development, and the strengthening of the competitiveness and implementation of development potential of all parts of the country. However, the total development is not possible if there are areas whose development is lagging behind the national average. For the purpose of identifying such areas, the act provides continuous assessment and classification of local and regional governments according to their level of development, and at intervals of 5 years. The I. group includes local government units whose development index value is lower than 50 % of the Croatian average in the II. group are those whose value ranges between 50 and 75 % of the average, in the III. group are those whose value is between 75 and 100 % of the average, in the IV. group are those with a value of 100–125 % of the average, and in the V. group are those whose value reaches more than 125 % of the average. Among them, the I. group and II. group receive the status of assisted areas.

This decision is issued by the government on the proposal of the Ministry of Regional Development and EU Funds (hereinafter: the Ministry), based on which it forms the program of sustainable social and economic development of assisted areas. The program defines the measures and projects promoting development, holders of their implementation, implementation activities, methods of monitoring and evaluation of individual measures, and amounts of funds for the implementation which are provided, as a rule, from the state budget. The most important measure to strengthen the fiscal capacity of local government units is the payment of state aid in the amount of a certain percentage of revenue from income taxes realized on the territory of the local units.[22]

In addition, the act contains a number of provisions about the documents that represent a starting point for strategic planning of the development of sub-national levels of government. This refers to the county development strategies, strategies for the development of urban areas, program for the promotion of regional competitiveness and urban development, and the program for sustainable social and economic development of assisted areas and other areas with development peculiarities. The Ministry, the Regional Development Council, and the Agency for Regional Development are the competent authorities for the coordination of activities of strategic planning at the central level; at the regions level, NUTS 2 partnership councils of statistical regions and partnership councils of County Associations; at the county level, partnership councils for the area of counties and urban areas and regional development agencies; and, finally, at the local level, local development agencies, and local action groups.[23]

Thirdly, from the recommendation of European institutions, it may be concluded that strategic planning[24] and participatory governance[25] are the best methods for achieving economic prosperity, social balance, and a healthy environment. During the 1990s, the creation of strategic development documents commenced in the Republic of Croatia and nongovernmental organizations and consulting companies was organized that offered a variety of services in the context of the preparation of planning documents. Neither did a legal obligation exist to prepare such documents,

---

[22]In the first year of the application of the status of assisted areas, aid shall be paid in the amount of 30 % of revenues from income tax, in the second year 50 %, in the third year 75 %, in the fourth and each subsequent year 100 % of the amount.

[23]The Ministry is, as the central state authority, the holder of the regional development policy. The Regional Development Council and the Partnership Council are composed of representatives of the ministries, regional and local governments, scientific and professional communities, businesses, social partners, and civil society organizations. They are responsible for reviewing development documents, proposing policy measures and programs, and monitoring their impact. Development agencies are established as public institutions or companies. Their main task is to encourage local and regional development, networking and cooperation of sub-national units, and the exchange of experiences and best practices among them.

[24]Theorists offer different definitions of strategic planning. Barry looks at strategic planning as a process of determining the objectives which an organization should achieve in the future and the measures to achieve them (Barry 1997). Pfeiffer et al. (1986), as well as Rothwell (1986) define strategic planning in a similar vein, as a process in which the managers of an organization plan its future activities and elaborate the necessary procedures and operations for the realization of the plans. Bryson (2011: 7–8), however, constructed the following definition: "Deliberative, disciplined approach to producing fundamental decisions and actions that shape and guide what an organization (or other entity) is, what it does, and why."

[25]The main characteristic of the participatory approach is the active involvement of all stakeholders in the development and implementation of development strategies. On the genealogy of the concept, see Stubbs (2009).

nor were the regulations for their production prescribed.[26] The issue of strategic planning begins to formalize in the last half of the decade, especially the Regional Development Act (*v. supra*) and the Budget Act, which obligates the administrative authorities to create a plan of development programs for these units for a 3-year period based on the strategic documents for the development of local and regional governments.

Strategic planning in the context of the public and therefore local government calls for a collaboration and coordination of relevant stakeholders, particularly in the context of sustainable development.[27] In a participatory democracy, the emphasis is placed on the "bottom-up" approach since the decision-making should 'descend' to the level of local people.[28] Its purpose is to increase the efficiency of decision-making and to increase the awareness of the problems of the local community. In addition, there are a number of measures that local authorities can use to support the development of entrepreneurship in their area. Puljiz (2005) states the following: (1) maintaining regular contacts with local entrepreneurs (local authorities so that they can get information on the needs of entrepreneurs), (2) finding and organizing sites and objects for businesses, (3) assistance to entrepreneurs in obtaining the necessary permits and approvals, (4) technical support, (5) tips about drafting measures to attract investments, (6) encouraging the development of formal and informal business networks, (7) raising the level of entrepreneurial culture (e.g., assigning a reward for successful entrepreneurship and initiating media campaigns), and so on. Local authorities should take into account such measures when designing development strategies.

Local development documents are usually directed toward an integrated development of the economy, society, institutions, and environmental protection. As a

---

[26]For this reason, the adopted development strategies were significantly different in content, structure, and scope. From 1991 to 2004, more than 100 national documents were published, which contained in their title the words "strategy," "plan," or "program," and were made mainly for formal reasons, and not because of the actual needs for a development document (Đokić et al. 2010).

[27]The concept of sustainable development encompasses the economic, social, and environmental dimensions. The first assumes the achievement of growth, efficiency, and equitable distribution of wealth (growth of agriculture and industry, efficient use of labor, satisfying the needs of households); the second relates to the participation in decision-making, mobility and cohesion, the realization of social identity, the development of institutions, and the like. The third dimension stresses the importance of integration of different ecosystems and the protection of the environment, natural resources, and biodiversity (Kordej De-Villa 2009; Sumpor 2009).

[28]It is possible to distinguish between the horizontal and vertical participation. The first form assumes the participation of different sectors, social groups, producers, companies, and so on, while the vertical participation refers to the hierarchy between national and local authorities in the decision-making process. Thus, the actors in the participatory process can be representatives of local and central governments, local people, civil society organizations, the private sector, research institutions, sponsors, and so on (Đokić et al. 2010). In theory, shareholders, it is possible to distinguish eight types of stakeholders based on three attributes (power, legitimacy, and emergencies): inactive, discretionary, demanding, dominant, harmful, dependent, critical, and 'non-stakeholder'. For more details, see Mitchell et al. 1997.

rule, they include an analytical part, results of SWOT analyses, and objectives as formulated solutions to problems with a separate group of goals with the highest level of priority. In a separate part of the document, measures and activities are elaborated with appropriate financial means for the implementation and respective authorities responsible for the implementation of certain measures, i.e., activities. In this sense, the development strategy is an instrument of transparency and optimum allocation of available resources. The institutionalization of relations between the European Union and the Croatian local development planning is increasingly gaining in importance as it opens the possibilities for resource mobilization from the European cohesion and structural funds.[29]

Fourthly, the principles of integrated urban development identified in the national legal framework can also be classified according to the dimensions of the Europeanization of domestic structures.

The institutes of the autonomy of local self-governments, functional and fiscal decentralization, solidarity/mutual assistance, altered relationships between the state–local government units, and established new institutions (development agencies, partner councils) are implemented in the dimension of political structures. The dimension of public policies has largely been Europeanized due to performed functional and structural adjustments. Mechanisms of strategic planning, consultation and dialogue, partnerships of local authorities with the private sector and civil society organizations, as well as the practice of twinning, multicultural integration, free public access to official documents of local authorities, principles of equality, and economic, social, and territorial cohesion have been adopted. Finally, in the dimension of political processes, institutes of local elections, local referendums, citizens' meetings/local committees, sending petitions, complaints and proposals of citizens, and so on have been implemented.

6.6 Conclusion

The subject of this research was downward causality in the area of European policies of urban management. It was attempted to provide answers to the two key research questions: What the principles and standards proclaimed by European policies on urban management are and whether the Croatian legal framework is in line with the recommendations of the EU and the CoE? Through the analysis of European documents, it was found that most of the recommendations are aimed at improving the quality of life in the cities, environmental protection, economic development and social justice, protection of monuments and cultural heritage, functional and fiscal decentralization, citizens' participation in local and urban

[29]Research on the Europeanization of local government units, which was conducted in January 2014, showed that each of the 30 surveyed cities on the Croatian territory, either alone or in partnership, has competed for funds from the European funds. Also, all of the surveyed cities have experience with creating a local development strategy. For more information see Dobrić (2014).

governance, achieving cooperation of local government units, the public and the private sector in their area, as well as the economic, social, and territorial cohesion. Strategic planning and participatory democracy were thereby identified as the most appropriate methods within urban governance processes for achieving sustainable growth and development, economic prosperity, social balance, and environmental protection.

Croatia replied successfully to the adaptation pressures by implementing the established catalog of principles and standards of integrated urban development. Although the majority of recommendations are implemented in the Constitution of the Republic of Croatia, in the laws and bylaws, there remains room for improvement, both at the national and local level. The government should primarily entrust local self-governments with a greater responsibility in the planning and implementation of national and European policies (including development policies and policies of urban management). Also, it is necessary to regulate the framework for the involvement of citizens in decision-making processes of local public policies and current issues. On the other hand, local authorities are encouraged to adopt a number of additional measures, such as: (1) to encourage the creation of advisory bodies (councils/boards) of foreigners at the local level, (2) to create conditions for the implementation of e-governance, e-democracy, as well as new information and communications technologies, (3) to create an effective system of providing information to the public, (4) to develop practices and mechanisms for evaluating the service delivery by local authorities, (5) to strengthen the coordination between related public affairs, and (6) as much as possible to include the private and public sector in the design and decision-making on local public policies.

Since research results have shown that Croatian cities and municipalities increasingly conduct functional (strategic planning, competition for resources from EU funds—alone or in a partnership) and structural reforms (development agencies, departments/offices or jobs for European issues, memberships in European institutions, and associations/networks of local authorities), the initial hypothesis is considered to be confirmed.

Finally, the relevance of the research results is twofold. The first part of the research could be used as a basis for researches of other EU and CoE Member States dealing with urban management because it has resulted in putting together a catalogue of the European principles for good local governance. The second part of the research represents a contribution to the interpretation of the Croatian legal framework for local self-government and development policies, since these issues have not yet been systematically analyzed. Also, the results have opened the possibility for further, empirical research of the identified European principles and standards in the sense of their practical implementation. This could be done by analyzing the national and European jurisprudence, administrative bodies' decisions, and quantitative and qualitative questioning of citizens' opinions.

References

Alexander L (2013) What are principles, and do they exist? Legal studies research paper series, No. 13-119

Alexy R, Rivers J (2010) A theory of constitutional rights. Oxford University Press, New York

Allen D (2000) Cohesion and structural funds. Transfers and trade-offs. In: Wallace H, Wallace W (eds) Policy-making in the European Union, 4th edn. Oxford University Press, Oxford

Bache I (2007) Cohesion policy. In: Graziano P, Vink MP (eds) Europeanization. New research agendas. Palgrave Macmillan, Hampshire, New York

Barry BW (1997) Strategic planning workbook for public and nonprofit organizations. Amherst Wilder Foundation, St. Paul

Börze T, Risse T (2003) Conceptualizing the domestic impact of Europe. In: Fetherstone K, Radaelli CM (eds) The politics of Europeanization. Oxford University Press, Oxford

Bryson JM (2011) Strategic planning for public and nonprofit organizations. In: A guide to strengthening and sustaining organizational achievement, 4th edn. Jossey-Bass, San Francisco

Dobrić D (2014) Europeanization of local self-governance. Dissertation, University of Zagreb

Dworkin R (2003) Shvaćanje prava ozbiljno. Translation: Rusan R, Matulović M. Zagreb, Kruzak

Đokić I, Rašić Bakarić I, Šišinački J (2010) Strategic program of local development—Croatian experience. Institute of Economics, Zagreb

Kassim H (2005) The Europeanization of member state institutions. In: Bulmer S, Lequesne C (eds) The member states of European Union. Oxford University Press, Oxford

Katenhusen I, Lamping W (2003) Demokratien in Europa. Leske+Budrich, Opladen

Kearney C (2013) Development programming, negotiation and evaluation: lessons for the future. In: Bachtler J, Turok I (eds) The coherence of EU regional policy. Routledge, New York

Kordej De-Villa Ž (2009) On sustainable development. In: Kordej De-Villa Ž, Stubbs P, Sumpor M (eds) Participatory governance for sustainable development. Institute of Economics, Zagreb

Ladrech R (2002) Europeanization and political parties: towards a framework for analysis. Party Polit 8(4):389–403

Ladrech R (2005) The Europeanization of interest groups and political parties. In: Bulmer S, Lequesne C (eds) The member states of the European Union. Oxford University Press, Oxford

Ladrech R (2009) Europeanization and political parties. Living Rev Eur Gov 4(1). http://www.livingreviews.org/lreg-2009-1. Accessed 1 Apr 2015

Ladrech R (2010) Europeanization and national politics. Palgrave Macmillan, Hampshire, New York

Mair P (2007) Political parties and party systems. In: Graziano P, Vink MP (eds) Europeanization. New research agenda. Palgrave Macmillan, Houndmills

McCormick J (2011) Understanding the European Union: a concise introduction. 5th edition. Palgrave Macmillan, Hampshire, New York

Mitchell R, Agle B, Wood D (1997) Towards a theory of stakeholder identification and salience: defining the principle of who and what really counts. Acad Manag Rev 22(4):853–886

Pfeiffer JW, Goodstein LD, Nolan TM (1986) Applied strategic planning: a how to do it guide. University Associates, San Diego

Puljiz J (2005) Economic development. In: Rüdiger P (ed) How to encourage development at the local level. Handbook with examples of the best practice from South-Eastern Europe. Zagreb, Friedrich Ebert Stiftung

Rothwell WJ (1986) The strategic planning workshop. Human Resources Development Press, Pennsylvania

Schlag PJ (1985) Rules and standards. UCLA Law Rev 33(2):379–406

Senden L (2004) Soft law in European community law. Oxford University Press, Oxford

Senden L (2005) Soft law, self-regulation and co-regulation in European law: where do they meet? Electron J Comp Law 9(1). http://www.ejcl.org/. Accessed 25 Mar 2015

Shapiro SJ (2007) The "Hart-Dworkin" debate: a short guide for the perplexed. Public law and legal theory working paper series, Working Paper No. 77

Sošić M (2006) Europeizacija nacionalnih politika: koncept i istraživački pristup. Anali hrvat politol druš 3(1):229–246
Sumpor M (2009) Introduction to development management, sustainable development and coordination. In: Kordej De-Villa Ž, Stubbs P, Sumpor M (eds) Participatory governance for sustainable development. Institute of Economics, Zagreb
Stubbs P (2009) Participation, partnership and/or support: internal and external stakeholders in sustainable development. In: Kordej De-Villa Ž, Stubbs P, Sumpor M (eds) Participatory governance for sustainable development. The Institute of Economics, Zagreb

Legal Documents

Council of Europe
European Committee on Local and Regional Democracy (2007) The Valencia Declaration. *MCL-15 (2007) 5 final, 16 October 2007*
The Congress (1992) European Urban Charter. *18 March 1992*
The Congress (1995) Opinion 1 on the draft Recommendation No. R (95) of the Committee of Ministers to Member States on the implementation of the principle of subsidiarity prepared by the CDLR. *30 May 1995*
The Congress (1996) Recommendation 19 on aspects of urban policies in Europe. *05 July 1996*
The Congress (2003) Recommendation 133 on management of capital cities. *22 May 2003*
The Congress (2008) Recommendation 251 European Urban Charter II—Manifesto for a new urbanity. *28 May 2008*

European Union

Bristol Accord, the Informal Council of Ministers on sustainable communities, Bristol, 6–7 December 2005
Committee of Regions (2002) Opinion on "Partnerships between local and regional authorities and social economy organizations: contribution to employment, local development and social cohesion". *OJ C 192, 12 August 2002*
Committee of Regions (2013) Opinion on "Empowering local authorities in partner countries for enhanced governance and more effective development outcomes". *OJ C 356, 05 December 2013*
European Commission (2001) European governance. A White Paper. *COM (2001) 428 final, 25 July 2001*
European Commission (2005) Communication from the Commission to the Spring European Council, Working together for growth and jobs: a new start for the Lisbon Strategy. *COM (2005) 0024 final, 02 February 2005*
European Commission (2008) Local authorities: actors for development. *COM (2008) 0626 final, 08 October2008*
European Commission (2010a) Europe 2020. A strategy for smart, sustainable and inclusive growth. *COM (2010) 2020 final, 03 March 2010*
European Commission (2010b) EU development policy in support of inclusive growth and sustainable development. Increasing the impact of EU development policy. A Green Paper. *COM (2010) 629 final, 10 November 2010*
European Commission (2013) Empowering Local Authorities in partner countries for enhanced governance and more effective development outcomes. *COM (2013) 0280 final, 15 May 2013*
European Parliament (2007) Resolution on local authorities and development cooperation. *OJ C 301E, 13.12.2007*

European Parliament (2010) Resolution on the draft Commission decision establishing the 2009 Annual Action Programme for Non-State Actors and Local Authorities in Development. *OJ C 212E, 05 August 2010*

European Union (2012) The Treaty on the Functioning of the European Union. *OJ C 326, 26 October 2012*

Leipzig Charter on Sustainable European Cities, the Informal Council of Ministers responsible for spatial planning and urban development, Leipzig, 24–25 May 2007

Lille Action Programme (2000) The informal council of ministers responsible for urban affairs, Lille. Accessed 3 Nov 2000

Urban Acquis, the Informal Council of Ministers responsible for territorial cohesion, Rotterdam, 29 November 2004

Republic of Croatia

Financing of Local and Regional Government Units Act, Official Gazette, no. 117/93, 69/97, 33/00, 73/00, 127/00, 59/01, 107/01, 117/01, 150/02, 147/03, 132/06, 26/07, 73/08, 25/12

City of Zagreb Act, Official Gazette, no. 62/01, 125/08, 36/09

Local and Regional Self-Government Act, Official Gazette, no. 19/13

Access to Information Act, Official Gazette, no. 25/13

Budget Act, Official Gazette, no. 87/08, 136/12

Regional Development Act, Official Gazette, no. 153/09

Environment protection Act, Official Gazette, no. 80/13

Public-Private Partnership Act, Official Gazette, no. 78/12, 152/14

Local Elections Act, Official Gazette, no. 144/12

Referendum Act, Official Gazette, no. 33/96, 92/01, 44/06, 58/06, 69/07, 38/09

Act on the Use of EU Structural Instruments, Official Gazette, no. 78/12, 143/13, 157/13

Chapter 7
The Relationship Between the European Commission and Local Government Through European Urban Initiatives: Constraints and Solidarities

Oussama Kharchi

Abstract The construction of the European Union in the last few decades piled up sectorial policies without taking into account the territorial impact of these policies. With regard to cities, even though they were the first to undergo the economic, social, and physical costs of the integration process, we had to wait until the drafting of the Single Act in 1987 and the structural funds reform of 1988 to see them emerging on the European stage. The recognition of cities as the relevant place for intervention is part of a wider reshaping process of Community action means. A reshaping process that intends, among other objectives, to establish a direct relationship between the European Commission and local governments in Europe. The purpose of this chapter was to put into perspective the European urban initiatives, in order to understand the relationship between the European Commission and Local Government and through this relationship to shed new light on the changing processes of urban governance in Europe. It will also expose urban problems and determine the capacity of city government in European countries to cross their strategies and needs with the objectives of European Commission. The chapter also aims to demonstrate how some European Commission exchange programs help to erase the political and administrative boundaries that are supposed to limit the competences of each territorial layer, while at the same time, they reveal new forms of collective learning by which the European Commission "Europeanize" knowledge and seeks to assert itself as a body of expertise for public action. This study is based on an evaluation research conducted in some beneficiary cities of European funds.

Keywords European Commission · Local government · Europeanization of urban policy · EU urban initiatives · UPP · URBAN · URBACT · Urban governance · European Union

---

O. Kharchi (✉)
Université Sétif 1, Sétif, Algeria
e-mail: oussama.kharchi@univ-setif.dz

© Springer International Publishing Switzerland 2017
C. Nunes Silva and J. Buček (eds.), *Local Government and Urban Governance in Europe*, The Urban Book Series, DOI 10.1007/978-3-319-43979-2_7

7.1 Introduction

For years, the attention of researchers interested in European integration was limited to explain, describe, and theorize the European construction, up to predict the future of the European Community. Since the implementation of the Single Market in 1992, researchers have focused on the functioning of the European Union and its policies. Hassenteufel and Surel (2000) distinguish three types of approaches of EU policies: the approach which examines action co-production by different levels of government; one that promotes the implementation of Community policies; one that is interested in the European Union as a place of production of Community policies. As for us, we will put more emphasis on the action co-production without neglecting the implementation of European urban initiatives.

A dense scientific production is available on the institutions of the European Union and European regions. The literature on cities and urban issues in the European Union has not experienced the same development. In recent years, studies were especially interested in cities as collective actors in a neo-Weberian sense (Häußermann and Haila 2004; Le Galès 2003). However, urban policies do not attract much attention. This deficiency is due, firstly, to the absence of explicit Community competence in the urban area and on the other hand, governments of Member States and European institutions, except the European Commission, were always more concerned about regional problems. Hence, the interest of a research focuses on European Urban Initiatives and the relationship between the European Commission and Local Governments.

First, we will put into perspective the interest of European Commission in urban issues, than a presentation of the first European Urban Initiative, namely Urban Pilot Projects, and its potential impacts on urban governance. Through the example of Lyon, the chapter will examine urban problems and determine the capacity of local government to cross their strategies and needs with the objectives of the European Commission. The next section will highlight the constraints in the implementation of EU programs. Last but not least, our ambition is to understand the process through which the European Commission is trying to act on urban knowledge and assert itself as a body of expertise for public action.

7.2 Urban Pilot Projects: Lyon Experience

The interest of the European Commission in urban problems began with the entry of the UK and Ireland into the European Community in 1973. These two countries, which had more important regional economic problems than those found in the other Member States—with the exception of Italy—had requested, among other things, the establishment of a financial instrument dedicated to regional development as a prerequisite for membership. The beginning of deindustrialization observed in these countries and elsewhere in the Community highlighted the urban question. The European Commission seized upon the problem, introducing a specific chapter in its first action report on the environment for the period of 1973–1976. This report pointed out that some problems related to urbanization and the geographical distribution of populations and activities in the countries of the Community would be treated more effectively at the Community level.

A milestone was reached with the enlargement of the Community to Greece, Spain, and Portugal. The economic situation in these countries pushed the European Commission to rethink macroeconomic development tools. This resulted in a de facto redefinition of objectives and resources of the European integration process, especially since the spillover dynamics seemed to be blocked (George 1993).

All these factors contribute to the reform of institutions (Smith 1996). Initiated by Jacques Delors, the reform sought to restore interest in "The European idea". A new target was introduced, namely economic and social cohesion. It was also necessary to improve the efficiency of decision-making and democratize it. In order to convince reluctant countries, the European Commission recalls that it acts according to the guidelines of the EU Council and the European Commission is best placed to ensure the common interest versus the particular interests of Member States. A management embellishing in the name of the effectiveness of Community action, which hides a rebalancing of relationships within the institutional triangle—European Council, European Commission, European Parliament—in favor of the European Commission.[1]

If the European Commission had substantial power of initiative and production of regulatory standards, it remained devoid of real means of implementation and its action did not cover all sectors. The 1988 reform enshrined the right of the European Commission to intervene directly in the implementation of structural policies. In substance, the European Commission could "on its own initiative and in accordance with the procedures provided for in Title VIII, decide to propose to the Member States that they submit applications for assistance with respect to measures of significant interest to the Community" (JOCE 1988a). Following Article 10 of the ERDF,[2] the European Commission co-financed two types of action: on the one hand studies and research and on the other hand pilot projects and innovative measures of economic development. There was no direct mention of the urban

[1]In his inaugural speech to the European parliament, on January 14, 1985, Jacques Delors said: "The European Commission must play a central role in engineering the construction of Europe."

[2]Article 10 of the new Regulation defines the eligible actions of the ERDF financial support: (a) studies at the Commission's initiative aiming to identify: the spatial consequences of measures planned by the national authorities, particularly major infrastructures, when their effects go beyond national boundaries; measures aiming to correct specific problems of the border regions within and outside the Community; the elements necessary to establish a prospective outline of the utilization of Community territory. (b) pilot schemes which constitute incentives to the creation of infrastructure, investment in firms, and other specific measures having a marked Community interest, in particular in the border regions within and outside the Community; encourage the pooling of experience and development cooperation between different Community regions, and innovative measures.

question or cities in the article.[3] Indeed, the European Commission, knowing the reluctance of some Member States with regard to the expansion of its areas of intervention, preferred to speak of "building infrastructure" with a "marked Community interest" and "innovative actions" while referring to the regional and not the city level. Moreover, the European Commission relied on the absence of economic eligibility indicators to the Funds of Article 10 to ensure the support of northern Member States. These countries, net contributors, favored any action likely to increase funding to which they were entitled. It is in this framework that the Urban Pilot Projects (UPP) were initiated in 1990, on an experimental basis, in the cities of London and Marseille, in order to explore new approaches to tackle urban problems. Since then, and under Article 10 of EEC Regulation No. 4254/88224, the European Commission has co-financed thirty-one other UPPs, in 11 of the 12 countries then in the EEC. The selection of the beneficiary cities was made by mutual agreement, little formalized, following the example of the UPP of Lyon.

After an informal meeting between the European Commissioner for Regional Policy (Bruce Millan) and the Mayor of Lyon (Michel Noir), in March 6, 1991, the European Commissioner informed the Mayor of the existence of Community initiative Urban Pilot Projects. He encouraged him to submit a project, guaranteeing him an ERDF allocation. On March 12, 1991, the Mayor passes along files on initiatives undertaken by the City—Success at School, On-site Action Plan, and Social Development of Districts (RÉGION RHÔNE-ALPES 1991)—to the European Commissioner. Subsequently, the European Commissioner sent a letter, on May 9, 1991, to the mayor, in which he expressed his interest in a project that would be presented by Lyon and would in particular aim to strengthen the links between schools and business. After an exchange between both entities, a definitive project is submitted by the City of Lyon to the European Commission on November 13, 1991. The example of Lyon also sheds light on the difficulties encountered in the implementation of the UPP Community Initiative.

Under the terms of ERDF Article 10, La Duchère,[4] a neighborhood in the 9th arrondissement of Lyon, benefited from European funding by decision of ERDF number 91/03/29/006, of December 9, 1991, for the realization of an UPP. The

[3]All policies of the European Union have some sort of impact on European cities, from the allocation of financial resources to the detailed regulations (environment, competition, etc.) that affect the structural context in which local governments and economic players act. In all, more than 40 programs and lines of action directly or indirectly affect the cities of the Union. Most of these actions were located in cities, but have not been developed for cities.

[4]Built in the 1960s on one of the three hills of the city, the district of La Duchère is 10 km from the city center and includes 5300 social housing and condominium units. Due to the development of instability, unemployment, and delinquency, La Duchère benefited from the Social Development of Districts policy. The first agreement between the State and the City, in 1986, allowed the initiation of a strong local dynamic in terms of rehabilitation (nearly 2700 units), equipment, but also academic success, insertion, and training of the at-risk public. Over the years and through the measures, the scale and fields of action are going to expand, but social and professional integration remains the common thread.

allocation (1.9 million €) represents 50 % of the predicted cost of the project. The UPP of Lyon La Duchère wanted to make "a contribution to the problem of upgrading the peripheral urban districts of social housing which pauperize, withdraw on themselves and become marginalized, by demonstrating the impact that an active development of the scientific and technical culture can have to stimulate the people to qualify professionally and to enhance the value of the district in the city" (Commission Européenne 1992). To do this, five objectives were established: academic achievement; orientation toward qualifications for access to employment; social and cultural integration; economic insertion of the most marginalized people; economic development of La Duchère and its integration in the city. These five objectives are related to five actions[5]: the construction of a discovery center; the creation of the house of employment, training, and enterprise; the establishment of a local employment structure; the construction of workshops for artisans and artists; and project promotion and exchange of experiences with other European cities.

To facilitate the involvement of institutional partners of the project—State, Region, County—and locate funding outside the ERDF, the authorities of the City of Lyon set up a steering committee whose mission was as follows: implementation of the project, its funding, its administration, and its financial management. This steering committee was co-chaired by the Mayor of Lyon and the Prefect of the region to put an end to the tension between the City of Lyon and the Prefecture of the region, each party wanting to lead the project. The Prefect, in a letter to the Mayor underlines that it is "contrary to the rules of the State to establish direct relations between the European Commission and the City of Lyon." In France, only the Prefect of the region handled the implementation of European policies and programs. However, the project design was done by the municipality without the participation of the Prefecture of the region. The latter made it clear, during a meeting (Comité de pilotage 1992), that the State had no plans to put in the UPP, of which it was kept little informed, additional financial resources. This tension between the Prefecture of the region and the City has complicated a little more the management of the project. Indeed, the project management chain—transit of funds and information—involved complex financial and administrative circuits, sources of conflict between the actors, and delays in the allocation of funds.

Locally, the mobilization of private partners and inhabitants was below the stated goals. Indeed, the involvement of associative structures and inhabitants did not exceed the data-gathering stage.[6] The reluctance of having to manage the involvement of associations, perceived as too reactive to any municipal project,

[5]Compared to the initial proposals submitted to the European Commission, two actions were abandoned: cultural identity and integration; cultural identity and urban art.

[6]The participation of the inhabitants did not exceed the data collection stage in most UPPs, with the exception of Cork, Dublin, London, Paisley, Rotterdam, and Stoke-on-Trent where the inhabitants were stakeholders throughout the implementation. These disparities in the participation of the citizens reflect national traditions. In England, the communities alone lead specific programs. The purpose of this involvement is to reduce the gap between global regeneration policies and local issues. In some cases, obtaining central government funding is subject to the participation of the inhabitants.

was the main motivation. Furthermore, these associations have expressed their opposition to the UPP, and acts of vandalism have targeted the local employment structure shortly after its opening as well as the Discovery Centre from the beginning of its construction. Regarding the lack of financial commitment from private partners, it was due to the adoption of a "ticket office logic" by the City toward solicited businesses. The latter did not want to be confined to the simple role of sponsors, but expected roles of full actors with their own goals. As a result, the financial participation of private partners, or rather of an informal group of industrialists and business managers, did not exceed 4 %[7] of the total cost of the project and was limited to the Discovery Centre. Doubts have long remained as to the very possibility of a financial plan.[8] Finally, the diversity of resources required by each of the actions has brought about the intervention of a large number of partners. It was essentially an institutional partnership, which has expressed itself through an important dialogue at the level of the actions, but which was not realized at the project level. The Steering Committee in which these partners participated was more a place of information and validation than of exchanges and collective construction. Most of the partners associated with the UPP were there in a function of representation of a department—Education, Insertion, Social Affairs, and Economic Development—and not in a function of realization of a project as beneficial to the Community as for each of them.

As of February 19, 1993, 14 months after the decision of the European Commission to grant funds for Lyon UPP, the financial plan still did not distinguish in a precise manner what is acquired and what was put forward. Ultimately, the cost of UPP was 5.9 million €, instead of the 3.9 million € initially envisaged, and was completed after 4 years instead of two. A UPP completed more under the empirical mode than from a previously defined method. The authorities in charge of the UPP favored the operational aspect to the detriment of the global objectives—qualification of the inhabitants, the involvement of the companies, and the opening up of the district. These bodies argued that the implementation of a European project was a process of discovery and learning. Nevertheless, it was essentially the basic elaboration of the project which was the major obstacle in the realization of the UPP. The actions

[7]More than 2/3 of URBAN I programs have been modified due to the withdrawal of private investors. In total, the financial contribution of private actors in the URBAN program accounted for 8 % of the overall cost. The social aspect is dominate in the program, the probability of private partners making significant profits from their initial investment was low.

[8]In a note dated January 30, 1992, and addressed to the City Councilman for academic success as well as to the managing director of Social and Academic Affairs by the Social Affairs Division, we can read, "Therefore, it is urgent and essential, before going further in the implementation of this program, to study the very possibility of a financial plan. Indeed, it seems hardly possible to design the financing of this operation by the mere aggregation of different allocations for actions having more or less direct links with the Pilot Program."

then developed according to their own logic, without reference to or linkages with other actions, and to the detriment of the integrated approach, promoted in the project presented to the European Commission. For example, to achieve consistent and qualifying pathways of insertion, synergy between the House of the employment and the local employment structure was targeted but did not materialize by any action. The interest of the local employment structure resided in its ability to generate an ongoing process from the emergence of carriers of new business start-ups, at the House of Employment and the incubator until the stage of maturation symbolized by the exit from the incubator, not to mention the rapid attainment of benefits in terms of jobs in the La Duchère District.

As regards the innovation and the exchange of experience, two key aspects for the European Commission, the results were very disappointing. The innovation, asserted in the project presented by Lyon, was shaken by the confrontation with the logic of reality during the implementation, which sometimes "brought them down" on more "traditional" logic. This is the case for the House of employment where the concrete contribution of business managers did not continue beyond the 2 years of realization. Another example, the ability of the "Local employment structure" apparatus to generate a continuing process, from the emergence of carriers of projects through to exiting the business incubator, remained virtual. In the same way, the innovative will to establish an artisanal and artistic pole with high added value was drained away. Each action had, in theory, an innovative dimension. However, it is the interdisciplinarity of actions with regard to the impoverishment of La Duchère that was innovative in many respects. Yet, as we indicated before, the coherence of the UPP was put to the test by the operational aspects.

As for the exchange of experience component, offers from the cities of Belfast, Dublin, Salamanca, and Swansea were made to the City of Lyon, but were not pursued. The Mayor of Lyon wanted to communicate on the UPP only once the project had been achieved. Furthermore, the City made a commitment to organize a seminar on the UPP for the end of the year 1995. As of today, this seminar has not been held. Generally speaking, communication within the UPP was neglected, making the image of the project "fuzzy" to the point of giving the impression of a project revised downwards compared with the initial version. Also, the redefinition of the UPP has made the transfer of the Lyon experience much less relevant. Creating commercial and artisanal premises, building workshops and offices or opening an employment agency has no demonstrative effect susceptible to interest other European cities. In a way, the Lyon UPP has enabled reflection on the development of disadvantaged neighborhoods.

7.3 Constraints in the Implementation of UPP

The difficulties encountered in the Lyon UPP or in the other projects refer first of all to the absence of a call for proposals[9] in the procedure for selecting initiatives and to the basic preparation of the projects. The European Commission maintains that for the selection of initiatives "the focus has always been on identifying projects that would enable the selected area to assist the functioning and well-being of the city rather than to handle directly its social or housing problems" (Commission Européenne 1992). Yet, the action of most of the projects targeted small areas facing social problems. Only the six UPPs targeting the revitalization of historic city-centers were part of a dynamic that focused on the urban structure as a whole.

Also, more than half of the UPP beneficiary cities are located outside of Priority Objectives areas of regional policy and therefore were not eligible for funding by the ERDF, whose mission is to reduce regional disparities. Some cities are among the richest cities in the European Union—London, Copenhagen, Brussels, Madrid, Rotterdam, Berlin, Lyon, Antwerp, Toulouse, Bordeaux, and Venice. The most revealing example is that of Rotterdam, which received 2.63 million € from the European Commission (50 % of total cost) to revitalize and fight urban decay and poverty in the Kop Van Zuid district. In fact, this UPP is only the social component of a large operation in this area, the cost of which has reached 2.5 billion €. In the case of Brussels, the ERDF financed the activities of local associations, some of which were already receiving aid from the European Social Fund.

In addition, the few applications or projects presented[10] by the cities to the European Commission show a lot of inaccuracy as well as a brief preparation of operations to be carried out. Some requests made to the European Commission barely constitute a working hypothesis. In employment, rehabilitation, or renovation of buildings, no quantitative targets have been set. In half of the UPPs, the proposed actions have not been carried out, even though the project implementation period was doubled compared to the initial schedule. Thus, in Copenhagen, the main UPP action, achieving an urban Ecotechnology center that recycles waste from the construction industry, was abandoned because of noise pollution and fire

[9]The use of call for proposals procedure was applied during the same period for other initiatives benefiting from a much smaller Community financial contribution. For example, 18 studies on the conservation of European cities were funded for € 1.5 million while 33 UPPs benefited from € 101.3 million.

[10]If in the case of Lyon, a project was indeed deposited with the European Commission, a doubt remains for the other cities. All our attempts to access these files have been ignored by the European Commission. We only had access to seven final reports: Groningen, Liege, London, Neunkirchen, Paisley, Rotterdam, and Thessaloniki. Faced with the unjustifiable silence of the European Commission, we brought up the question with Mr. André Carneroli, Director of the Audit Group II—Structural and internal policies—at the European Court of Auditors. He suggested the hypothesis of the nonexistence of these files: "qu'il y a eu des documents présentés ou pas ça on ne le sait pas, j'ai plutôt l'impression qu'il y a eu une entente préalable pour savoir qu'est ce qu'il faut faire pour qu'il y est possibilité de financement". His team also did not have access to these files when preparing the Special Report on the urban environment (JOCE 1994).

risks due to the use of flammable materials. In the case of Genoa, the renovation and the upgrading of the acropolis of Genoa, as well as the conversion of Santo Salvatore church into a cultural center, were not realized. In Gibraltar, the development of centralized waste treatment facilities was not concretized. For Marseille, no action has materialized on the improvement of public transportation. Finally, in Aalborg, the urban renewal center closed its doors after being opened for 6 months.

According to the provisions governing the Structural Funds (JOCE 1988a, b), the Community aid, from which the cities of the UPPs have benefitted, should be forfeited in whole or in part if the nature or conditions of implementation of the projects experience significant changes. Although the quasi-majority of UPPs are in this case, the European Commission has only sanctioned the city of Thessaloniki by deducting € 250,000 of its initial allocation, which represents 4.5 % of the total amount (Commission Européenne 1997b).

The innovative action desired by the European Commission presupposes the idea of a new alternative to existing solutions and procedures. In reality, the measures designated jointly by the cities and the European Commission (Commission Européenne 1995; RECITE 1995) as innovative were similar to those already in practice. For Antwerp, innovation lies in "an interview between a counselor and client to identify a way forward and identify customer needs in terms of training and support." In Bilbao "self-rehabilitation is a major innovation." Another example, Stoke-on-Trent where "to provide opportunities for training and employment is an innovation" (Regional Policy 2014). All of these measures are similar to those implemented in most Member States and even in the operational programs on the renovation of historic centers or on the return to employment of objectives 1 and 2. One of the most common actions in employment and the fight against social exclusion was the creation of service centers, which propose information services, career guidance and/or training to the local population, as well as support services or counseling for small- and medium-sized enterprises. These mechanisms used in German, Belgian, British, French, and Dutch UPPs were already a current practice in those countries (Van Den Berg et al. 1998). UPP local managers consider that the sluggishness in European and national financial and administrative channels led to the renunciation of the rare innovative aspects of the projects. By giving the priority to the procedures at the expense of the process, the interest of European Commission initiatives in the urban field is much more limited in terms of new model development or acquisition of transferable knowledge.

This aspect partly explains the virtual absence of procedures for exchange of experiences between cities, apart from organizing conferences, and publishing information letters on the projects. Out of 33 UPPs, 11 have organized conferences and 3 have organized workshops. Mere coincidences, or perhaps a condition of the European Commission, the cities which have organized these meetings, with the notable exception of Athens and Dublin, are the last to have been chosen for the UPP program. The participation level at these seminars was 23 %, and it is practically the same cities that attend these meetings. To explain, the variable interpretation of the "exchange and transfer of experiences at the Community level" component, our interlocutors, in the cities, invoke difficulties with the language of communication and the differences in politico-administrative culture. This latter argument is consolidated by the establishment of networks of exchange, mainly on administrative procedures of the Structural Funds, between cities of the same Member States: Germany, Austria, France, Ireland, and the United Kingdom. Moreover, some UPP local managers have expressed willingness to share their experience, but "did not know what the European Commission wanted because nothing was asked from them." On the contrary, other managers did not see the interest in the exchange of experiences.

In light of these factors, we can argue that the establishment of the urban initiatives was a pretext for institutionalization of relations, up to that point informal, between local governments and the European Commission. Solving urban problems was not an end in itself, but a means to mobilize cities without the control of Member States. This first program developed by the European Commission can be regarded as display logic.[11] Ease of Community aid has driven a dynamic incentive to mobilize cities on the occasion of the launch of the second UPP series. Following the call for proposals of November 30, 1995, the European Commission received 503 proposals. If in the first series of UPP all cities had received Community subsidies, only 26 of 503 cities had acquired it in the second series after a drastic selection.[12] The second series of UPP perfectly illustrates the passage of "incentive" mobilization to "selective" mobilization. Initially, the European Commission was in search of recognition and direct dialogue with cities, and then, the cities began looking for European Commission "favors". Virtually, all EU cities over 250,000 inhabitants in 1995 had submitted a proposal in the framework of the second series of UPP.

Moreover, this incentive mobilization is behind the implementation of the URBAN program. According to Bruce Millan, European Commissioner for Regional Policy, the European Commission is simply responding to a growing demand from cities affected by the same problems in some of their neighborhoods (Cour des Comptes Européenne 2001). However, the granting of a Community allocation has become conditional and only the cities that develop projects in line with the European Commission agenda obtain it. Furthermore, UPP or URBAN, whose demonstrative value cannot be measured in terms of the Community allocation—on average € 2.8 million by UPP and € 8.7 million for URBAN—must be assessed in terms of their ability to change the way cities tackle urban problems. The European Commission

[11]Objective 2 had an urban component (aid to businesses, transport and communications infrastructure, etc.), with a budget of € 2 193 million for the period 2000–2006. According to European Commission estimation, nearly € 16 billion was spent in the cities under objectives 1 and 2 during the period 2000–2006. Training measures under Objective 3 and EQUAL covered urban areas in difficulty.

[12]One year of careful examination of the proposals by 27 independent experts commissioned by the European Commission. The evaluation of the experts focused on four points: the innovative character of the proposal at the national and EU level; the demonstration potential and applicability in other contexts; the partnership envisaged in the project; the impact of proposed actions on employment.

considers that the relative failure of local authorities to deal with urban problems is a matter of procedure (Commission Européenne 1997a). Thus, efforts were focused on the methodology and incentive virtues of obtaining EU funds. This is based on the results of the actions undertaken in UPP and URBAN programs. In more than two-thirds of cases, the programs have not achieved their objectives in terms of job creation. On the other hand, the procedures developed in the framework of these programs, especially partnership and the integrated approach, figure as reference in the Member States with little experience in urban policy.

7.4 Europeanization of Urban Policies: URBACT Program

This last point leads us to speak about the institution of the European Commission as a body of expertise on the stage of public action. European integration, according to Jean Monnet, establishes a special relationship between knowledge and public action. This is done by gathering currents of thought to challenge existing paradigms. The introduction of these new ways of thinking and doing is done through the exchange of experience programs and projects certified as "good practice" by the European Commission. The idea is ultimately to circulate and share values to make them common.

In the official brochure of Exchange of Experiences Program, which dates to 1990, it is mentioned that "the exchanges of experiences aim at encouraging communities to adopt, at their level, an European point of view, and thus to participate more actively in the implementation of Community policies, as well as to facilitate the dissemination of ideas and know-how between the decentralized communities."[13] In the urban field, the European Commission was confronted with the lack of interest of cities in the exchange of experiences, in initiatives such as UPP or URBAN, whether during the implementation of the projects or after their completion, and that even though it was a priority for the European Commission. The common interest, namely discovering other ways and to judge its own ways, was not enough to spontaneously generate collective action. For Olson (1965), collective action is impossible without distribution of individualized remunerations. The participation of cities remains dependent on support by the European Commission of the costs of cooperation. The European Commission had to set up a program with a specific budget (24.76 million €, 16 million from ERDF), URBACT, to establish an exchange between beneficiary cities of UPP and URBAN.

[13]The Exchange Experience Program comes under the budget line B2-600 and appears alongside regional action in the framework of Article 10 of the ERDF (Commission Des Communautés Européennes 1990).

The work of creating thematic exchange networks of URBACT program was relatively easy. However, the conversion process of the information held by each actor in useful and exploitable resources by all program partners was more difficult. On one hand, the "portability," which is one of the keywords of the Community discourse, is based on a proactive process of learning and appropriation of external knowledge. On the other hand, the implementation of UPP and URBAN, supports of the exchange, was made in different economic, social, and environmental contexts and with a particular vision of the role of the State which makes the comparisons and the transfer of practices difficult. The example of Bremen and Groningen UPPs illustrates well the question of the "localization" for the treatment of urban problems. Sharing the same objective, namely the socioeconomic reactivation of a district characterized by a strong proportion of unemployment, both cities had adopted a similar type of intervention and a set of actions with a joint collaboration in the implementation of the two projects. The relative success of Groningen[14] UPP contrasts with the failure of Bremen's.[15] Nearly 5 years after its approval by the European Commission, the URBACT program work led to the development of a knowledge transfer kit.[16] The purpose of this device is to "facilitate the transfer of know-how dispersed between different levels of government by helping the emergence of a capability of local expertise" (URBACT 2009). Let us recall that the initial objective of the URBACT program, according to the terms of the European Commission, was that "the exchanges are used to develop better integrated local policies and also in the design of national policies" and "to learn from the analysis of these experiences, and from policies led locally and to propose innovative approaches to address these difficulties" (JOCE 2000).

Under cover of a policy of consensual approach and open consultation, organizational mechanisms are set up to control mobilization and exchange.[17] In the ILE program, EEP, RECITE, DIRECTORIA, Open Days, or URBACT, the logic behind the action of the European Commission is not only the manifestation of a

---

[14]Named SEND (*Social Economic Network Development*), the project aspired to boost employment and economic regeneration in the district of Korreweg-Oosterpark. Two lines of action were carried out: the establishment of a local employment agency (25,000 visits a year) and the creation of a network of small- and medium-sized enterprises to favor occupational integration (48 jobs created for a total of 94 people who have received training).

[15]The aim of the project was the socioeconomic reactivation of a peripheral residential district, Osterholz-Tenever. A regional partnership agency and training center were established as well as the implementation of local services. All these actions failed. Business involvement in training schemes lasted only 1 year, and that, although the training was established according to their needs.

[16]It is a matter of organizing seminaries at the invitation of Member states or city associations. Each seminar is dedicated to a particular topic that has been the subject of exchange within the URBACT program with the support of documents in local language for decision makers and practitioners in cities and in the presence of URBACT experts and representatives of partner cities.

[17]A figure allows a better illustration of the scope of transnational cooperation: Two thousand projects were completed in thirteen areas of cooperation for the purpose of territorial development, between 1990 and 1998.

lack of knowledge that would indicate the difficulties related to the inability of local governments to solve urban problems. It is situated more broadly in a theoretical perspective of creating new knowledge to redefine the terms of the action and to disseminate and introduce new ways of thinking and doing. We are witnessing in some ways an Europeanization of knowledge.[18] Europeanization, in the sense of a strengthening of organizational ability of the European Commission as a level of action absolutely necessary to the local mobilization. Indeed, the proliferation of urban programs, under cover of the objective of economic and social cohesion, leads to the recognition of the local as a legitimate level of Community intervention. The European Commission wants to be a full actor, interacting with a maximum of interlocutors (Hooghe and Marks 2001). That is to put issues of partnership and subsidiarity at the center of debates on the process of European integration.

As a principle of programming, partnership is also a principle of political structuring and of the legitimacy of the actors in close relationships. It carries an integrating current in the direction of a constant expansion of the combined action of the European Commission and local government. However, it does not solve the problem of prerogatives assigned in each country, to the state and local authorities, for the implementation of Community policies. Part of the answer lies in the establishment of exchanges and transfer of experiences. The existence of these procedures can erase the political and administrative boundaries intended to govern the blocks of competence at every territorial level. Meanwhile, these procedures reveal new forms of collective learning by which the European Commission "Europeanizes" knowledge (Goldsmith 1993) and seeks to assert itself as a body of expertise on the stage of public action. Exchanges often pass through the associations of cities. The European Commission helped the creation of these associations and contributes to their maintenance by the allocation of funds for the implementation of cooperation projects or through study contracts and assistance. In return, the European Commission expects support from local governments for the adoption of its propositions by Member States. The proliferation of transnational networks of cities is a good indicator of the development of horizontal interactions.

Subsidiarity defines a mode of relations between the European Commission, Member States, and local authorities based on the priority given to the territorial level most relevant to carry out certain goals. Introduced by the Single Act, subsidiarity was limited to environmental issues before being extended to other areas with the adoption of the Maastricht Treaty. According to its Article 5, the European Commission is authorized to intervene in areas which do not fall within its exclusive powers, under the principle of subsidiarity, "only if and insofar as the objectives of the proposed action cannot be sufficiently achieved by the Member States and can therefore, by reason of the scale or effects of the proposed action, be

[18]Some Member States do not remain inactive. Thus, at the initiative of France and the Netherlands, a European Urban Knowledge Network was established in October 2005. It is a Web portal on urban policies in Europe. The idea is to rely on the various existing networks of cities and research as well as the URBACT and URBAN programs to connect European knowledge bases on urban issues.

better achieved at the Community level." Thus, the establishment of the UPP or URBAN program is not considered as necessity, but it is part of the effectiveness of the public action. These programs therefore fall under the Community added value and are in accordance with the principle of subsidiarity. They are designed as support programs and experimentation of local knowledge. The European Commission does not wish to substitute itself for States and Cities which are better placed to carry out most actions, but act in a complementary fashion to their actions. These precautions in the speech are aimed at certain Member States, which are wary of the increasing incursion of the European Commission in areas that are within their competence.

7.5 Conclusions

The action of the European Commission is not limited to a mere financial contribution. The European urban initiatives, such as UPP, URBAN and URBACT, are vectors of integration, a process through which the political, social, and economic dynamics of the European Union become part of local policies and of urban governance processes. This process, which concerns both individuals and institutions, is not mechanical. It is rather a set of transformations with different extent and nature. Disseminating and introducing new ways of thinking and doing became the main European Commission initiative in urban field.[19] If UPP and Urban programs have only known two phases, URBACT is in its third phase. The total eligible budget of URBACT III (2014–2020) is 96.3 million € (74.3 million ERDF contribution) that is about four times URBACT I budget (24.76 million €). The program covers all Member States and will develop three types of intervention: transnational exchange; capacity-building; and capitalization and dissemination. Projects financed under URBACT III will address four main objectives: improving the capacity of cities to deliver sustainable urban policies; improving the design of urban policies; improving policy implementation in cities; and building and sharing knowledge (European Commission 2014).

The extension of European Commission action means that legitimate and asymmetric dialogue between European Commission and central governments is replaced by a conversation with several participants. At first, the relationship between cities and the Commission was quite balanced and even favorable to Cities. The Commission, in search of legitimacy, needed political support and relay. To obtain the cooperation of cities, the Commission increased the financial incentives and showed great flexibility on the real Community interest of cities projects that have benefited from the Structural Funds. Subsequently, the criteria to

[19]INTERREG program had also an impact on disseminating and introducing new ways of "thinking" and "doing" on urban issues. The fifth programming period of Interreg has a budget of EUR 10.1 billion invested in over 100 cooperation programs between regions and territorial, social, and economic partners.

obtain structural funds become stricter and more complex than those of national programs. The relationship between Cities and the European Commission becomes more favorable to the latter.

The formation of a Community action space generates a redistribution of authority. The more the interdependencies between Member States grow, the more the coordination problems are reinforced and promote the extension of European Commission powers. In response, Member States agree to transfer powers to the Commission only if in return they control certain aspects of the delegated powers. In this relationship, central governments are sometimes supported by regional governments with extended powers—Germany, Austria, Belgium, and Spain—that express more and more their reluctance to the expansion of the European Commission's prerogatives which would signify a limitation of their field of action. Thus, in the field of regional policy, eligibility rules in the programming period 2007–2013 were solely defined at the national level rather than at EU level as in the past. In auditing, inspections can be carried out by national bodies, according to the applicable national regulations.[20] Another example, in URBAN II program, the Member States have also recovered the most important prerogative, namely the selection of beneficiary cities and the distribution of financial allocations. An operation that has achieved two goals simultaneously: first, to end the position of the European Commission as a direct funding source for cities, and secondly, to select the cities according to domestic political considerations, without considering the innovative aspect of the proposed approach.[21] To conclude, let us say that what is at stake is less the strict and rigid application of the subsidiarity principle than a spirit of cooperation between the various levels of government. In other words, urban governance in Europe should be seen within the multi-level governance in which the local executives are involved in the development of EU and national policies that affect their territories.

References

Commission des Communautés Européennes (1990) Info-technique, T-508. DG XVI, Bruxelles

Commission Européenne (1992) Projets Pilotes Urbains. DG XVI, Bruxelles

Commission Européenne (1995) Guide des actions novatrices du développement régional (article 10 du FEDER) 1995–1999. Office des publications officielles des Communautés européennes, Luxembourg

Commission Européenne (1997a) La question urbaine: orientations pour un débat européen, COM (97) 197 final. Bruxelles

[20]For programs whose budgets do not exceed € 750 million of which the Community contribution is a maximum of 40 %.

[21]For example, Sweden and France have favored sites already covered by national regeneration programs. For Germany and Belgium, the selection has resulted in a program by Lander state or region.

Commission Européenne (1997b) Projets Pilotes Urbains, Rapport Annuel 1996. Office des publications officiels des Communautés européennes, Luxembourg

Comité de pilotage (1992) Rapports des réunions 28/01/1992, 02/10/1992, 06/11/1992, 12/03/1993, 16/04/1993, 02/11/1993, 04/01/1994, 531 17/06/1994. Direction de l'action sociale, Ville de Lyon

Cour des Comptes Européenne (2001) Rapport spécial n° 1/2001 relatif à l'initiative communautaire URBAN, accompagné des réponses de la Commission. Journal officiel des Communautés européennes, C 124, 44e année, 25 avril 2001, Luxembourg

European Commission (2014) The urban dimension of EU policies – key features of an EU urban agenda, COM(2014) 490 final, Brussels

George S (1993) Politics and policy in the European Community. Oxforf University Press, 3rd edn

Goldsmith M (1993) The Europeanisation of local government. Urban Stud 30(4–5):683–690

Hassenteufel P, Surel Y (2000) Des politiques publiques comme les autres? Construction de l'objet et outils d'analyse des politiques européennes. Politique européenne 1(1):8–24

Häußermann H, Haila A (2004) The European city: a conceptual framework and normative project. In: Kazepov Y (ed) Cities of Europe: changing contexts, local arrangements and the challenge to urban cohesion. Blackwell, Oxford, pp 43–63

Hooghe L, Marks G (2001) Multi-Level Governance and European Integration. Rowman and Littlefield, Lanham

JOCE (1988a) L 185 du 15/07/1988:9-20. Office des publications officielles des Communautés européennes, Luxembourg

JOCE (1988b) L 374 du 31/12/1988:15-20. Office des publications officielles des Communautés européennes, Luxembourg

JOCE (1994) C 383 du 31/12/1994. Office des publications officielles des Communautés européennes, Luxembourg

JOCE (2000) C 141/04 du 19.05.2000. Office des publications officielles des Communautés européennes, Luxembourg

Le Galès P (2003) Le retour des villes européennes. Sociétés urbaines, mondialisation, gouvernement et gouvernance. Presse de Sciences-Po, Paris

Olson M (1965) The logic of Collective Action. Harvard University Press, Cambridge

RECITE (1995) Projets Pilotes Urbains, Bulletin 5, 6, 9, 10. Bruxelles

Regional Policy (2014) Urban Pilot Projects. http://ec.europa.eu/regional_policy/archive/urban2/urban/upp/src/frame1.htm. Accessed 8 April 2014

RÉGION RHÔNE-ALPES (1991) Charte d'objectifs, développement social du quartier de la Duchère, Contrat de Plan État-Région 1989–1993. Lyon

Smith A (1996) L'Europe au miroir du local. La réforme des Fonds structurels. L'Harmattan, Paris

URBACT (2009). http://www.urbact.eu. Accessed 8 Jan 2009

Van Den Berg L, Braun E, Van Der Meer J (1998) National Urban Policies in the European Union: Responses to urban issues in the fifteen member states. Ashgate, Aldershot

Chapter 8
Mega-Events and New Patterns of Cooperation: The European Capitals of Culture

Ágnes Németh

Abstract Since its launch 30 years ago, the European Capital of Culture (ECOC) programme has been researched widely from different perspectives of urban and regional development, investigating the physical and structural changes that it initiates and the challenges that it brings to local government and to urban governance. Alternatively, this chapter gives a longitudinal analysis of two particular ECOC cases to examine the potentials of this cultural mega-event for social (network) capital mobilisation. Inter-organisational networks of an ECOC project can be characterised as collaborative networks; however, considering their potential longer-term sustained impact, they can contribute to strengthening the local–regional governance network. Using examples of the cities of Pécs, in Hungary, and Turku, in Finland, the analysis looks into the networks and relations between various participating actors in order to explore how these ECOC projects facilitate and change inter-organisational cooperation and networking among cultural and creative producers. The study applies the basic approach and methodology of social network analysis and uses primary data taken from online surveys performed in the two ECOC cities. Primarily, the chapter uses an analytical approach based on the thorough examination of numerical and visualised data. The findings reveal the importance of knowledge and support networks in the field of creative industries and the role of large-scale cultural events in the formation and sustainability of such networks.

Keywords European capitals of culture · Local government · Urban governance · Mega-events · Collaborative networks · Local–regional development · Social network analysis · Hungary · Finland

---

Á. Németh (✉)
Karelian Institute, University of Eastern Finland, Joensuu, Finland
e-mail: agnes.nemeth@uef.fi

© Springer International Publishing Switzerland 2017

C. Nunes Silva and J. Buček (eds.), *Local Government and Urban Governance in Europe*, The Urban Book Series, DOI 10.1007/978-3-319-43979-2_8

8.1 Introduction

Over the last two decades, there has been an increasing interest in regional studies regarding the horizontal relations of local and regional actors, based on the awareness that networks play a key role in social, cultural and economic development. Collaborative networks are seen as important assets in various fields, from R&D practices (Krätke 2010), to urban cultural development (Comunian 2011a) and environmental protection (Hirschi 2010). Such networks have become part of regional development strategies,[1] mostly as promoters of innovation and economic competitiveness.[2] What is common to most of these case-study research and development strategies is that they see the processes of cooperation and networking among different types of actors as an effective and essential governing power.

In the case of mega-events (e.g. the Olympic Games, the Football World Cup, the World Expo or the European Capital of Culture), the development of inter-organisational cooperative networks that shape these events is based on a complex web of relationships that exist both inside and outside the event hosting community. This chapter engages with current discussion about the formation and role of networks in framing governance strategies, especially in event tourism contexts (Edizel 2014; Morellato and Williams 2014; Ziakas and Costa 2010). In particular, it places a special focus on the *complexities* and *dynamics* of cooperative networks. Using research undertaken in Hungary and Finland, this chapter investigates the networks and relations between various actors participating in the two European Capital of Culture projects (Pécs 2010 and Turku 2011). It aims to make both a theoretical and an empirical contribution to reach a better understanding of the network-oriented forms of urban governance, as well as to engage in debates about social and inter-organisational network analysis.

Previous studies and conceptual developments related to the relationships that mega-events and the collaborative networks of organisations have with local–regional development provide a background to the chapter and inspire some initial hypotheses for investigation. Thus, the empirical analyses of the two ECOC cases are preceded by a clarification of relevant concepts and typologies and also a brief review of the academic accounts of these relationships. As a result, some assumptions based on existing debates are instrumental in answering the initial questions raised in the chapter and are later tested in the analysis.

---

[1] In the European Union's Cohesion policy, it is referred to as a 'partnership'.

[2] Finland's Regional Development Strategy 2020, and to a lesser extent Hungary's National Development and Territorial Development Concept 2030.

8.2 Relevance and Hypotheses

8.2.1 Mega-Events and Social Networks

Mega-events are increasingly perceived as strategic tools for local and regional development (Evans 2011; Hall 2006; Mills and Rosentraub 2013). The anticipated and perceived benefits of holding such a major event are most often related to material and financial returns. However, it is just as important to consider how *large-scale events may help shape and strengthen a community and its social infrastructures.* In the absence of considerable economic benefits (which is often the case: see Palmer 2004), this latter, less tangible outcome may in fact represent one of the few advantages a city can gain from hosting a mega-event. Also, in the long term, the potentials for sustainable positive socio-economic impacts can be realised by the mobilisation of existing but often inactive resources through the formation and sustenance of collaborative networks. Therefore, an analysis of mega-events in terms of the stakeholder networks which potentially emerge, expand and operate during their implementation may help us to better understand the local–regional development implications of these events.

In comparison with tangible economic benefits (i.e. tourism flows and incomes or spectacular urban regeneration projects), the somewhat less tangible role of social capital and the emergence and behaviours of social networks are still relatively underrepresented in the field of mega-event research. The analysis of stakeholder networks and inter-organisational relationships has featured in academic recent discussion as one of the leveraging legacies of hosting mega-events, especially in the literature dedicated to major sporting events. O'Brien and Gardiner (2006) analyses the role of the Sydney Olympic Games in commercial development through the event's contribution to the creation and reinforcement of business relationships, while Misener and Mason (2007) have studied the significance of social capital in urban regeneration related to the 2002 Commonwealth Games in Manchester, UK.

In regard to cultural mega-events, academic discussion about social networks seems to be less active. Recently, stakeholder collaborations have been studied as related to festivals (Getz et al. 2007; Morellato and Williams 2014); however, the majority of ECOC case studies still have a strong focus on the event's economic and image-building impact (Gomes and Librero-Cano 2014; Herrero et al. 2006; Richards and Wilson 2004; Steiner et al. 2015). The consideration of how the ECOC would affect the networking of different stakeholders appeared only very recently in the evaluations (Richards 2015).

The bulk of relevant literature does not explicitly test the claim that networks are a better way to govern or that they are more efficient than hierarchies or other alternative means of organisation. However, many analyses share an instance that *innovation and sustainability may be important consequences of network governance.* Since innovation (raising the event's attractiveness) and sustainability (creating legacies for the intense financial investment) are both important expectations in the case of the European Capital of Culture programme (Németh 2010), it is interesting to investigate whether and how social networks are supportive of these claims. Furthermore, a collaborative network approach is even more relevant for studying the local impacts of the ECOC programme because in the case of both of the studied projects, the inclusion of collaborative acts (performances, exhibitions and other activities) was particularly encouraged in the initial calls (Németh 2015).

8.2.2 Social Networks and Local–Regional Development

Generally, a social network can be defined as a series of established relationships between interdependent actors within a larger social structure. The 'nodes' of networks can be individuals or groups, as well as formal organisations, thus making up so-called inter-organisational webs. The significance of *inter-organisational* linkages was already being discussed in the 1960s with initial focus placed on the relationships in-and-between business organisations. The academic literature of the last few decades mostly comes from the field of management studies and concentrates on the organisational relationships within different industries (Müller-Seitz 2012; Powell 1990). However, this kind of network approach has found its way into social-scientific studies which are interested in the impacts of networks on public service provision (e.g. health care), policymaking and governance (Ansell 2000; Bulkeley 2005; Provan and Milward 2001).

Isett et al. (2011: 161) distinguish between three separate public management networks: *policy, collaborative and governance networks*, where policy networks seek to shape public decision-making, collaborative networks work together in the delivery of services and governance networks combine aspects of policy making and service delivery. Inter-organisational networks which form and operate in the context of an ECOC project can be seen to be most similar to collaborative networks, yet horizontal collaborative linkages can also enrich the governance networks of the localities involved, their regions and local/regional development.

Concerning inter-organisational networks, the 'whys and whens' of network formation have been widely researched aspects in the context of emergent networks (Cropper et al. 2008; Harrison 2013; Kilduff and Tsai 2003). There are various drivers and purposes of network formation at work. A frequent driving force is resource dependence or resource sharing (combining), where one party is dependent on the resources and competencies which are controlled by another, where there are perceived or actual mutual gains, and where the synergies of providing resources and competencies are shared. Sometimes, these more internal motives are combined with an external pressure to 'team up', for example when exerted by the rules established by a funding authority. The inter-organisational cooperation networks involved in ECOC events have indications of both *internal and external* motives (Németh 2015).

Considering the *sustainability of networks*, Newman and Dale (2005) argue that the more heterogeneous and diverse a network is, the more resilient it will be to future changes. Along this line, they have further developed the concept of the utilisation and reproduction of social capital (first discussed by Pierre Bourdieu 1983/1986) by emphasising the significance of the combination of 'bonding' and 'bridging' ties (Putnam 2000; Woolcock 2001) within a network. In their analysis, networks composed of 'bridging ties' (connecting various types of actors with different resources, e.g. across economic sectors or fields of activities) are argued to strengthen a community's ability to adapt to change. 'Bonding ties' (connecting similar actors), however, create dense networks that may increase trust between members, but at the same time may also encourage either conformity or exclusivity, thus leaving less room for fresh ideas and experimentation. Accordingly, reaching and maintaining *a dynamic balance of the bridging and bonding types of links in a network* increases adaptive capacities and supports sustainability. In the case of an emergent network such as one reactivated and/or forming around an ECOC project, it is of significance whether this balance is created in a way that it will contribute to the continuity of cooperation and synergic effects in the city and its wider region, and also follow on from the actual event year.

Based on the arguments above, the chapter starts with the normative claim that inclusive horizontal cooperation among diverse stakeholders in the implementation of mega-events is beneficial. In more detail, it is assumed that:

- ECOC projects (and in general, large-scale events) may help shape and strengthen a community and its social infrastructures, which especially in the absence of considerable and more tangible economic benefits, can represent an important advantage that a city and its region can gain from hosting a mega-event.
- The innovation and sustainability of positive achievements from a mega-event may be significant consequences of network governance, both in the particular context of governing mega-events, such as the ECOC, and in local and regional development (policy making, service provision) in general.
- The heterogeneity of the ECOC networks and a balanced combination of bridging and bonding linkages support the continuity and sustenance of cooperation and its positive impacts on local–regional development.
- Inter-organisational networks operating towards the implementation of an ECOC project are probably most similar to emergent collaborative networks, but also, considering their longer-term sustained impact, they contribute to strengthening the local–regional governance network.

As implied above, the chapter investigates the networks and relations between various participating actors in order to explore how European Capital of Culture (ECOC) projects facilitate and alter the patterns of inter-organisational cooperation and networking among cultural and creative producers. Within this framework, this research addresses three more concrete questions:

1. What is the level of networking between the organisations who are involved in one way or another in the implementation of the featured ECOC projects, and are there any unique patterns which are observable in their cooperation networks?
2. To what extent do the activities related to the organisation of ECOC events initiate new and/or revive former relational spaces? In particular, do these relations build temporary structures (i.e. operating only for the ECOC event year) or are they capable of forming more permanent (sustained) networks which also function after the event?
3. How does the intensity of networking affect the experience and opinions regarding the benefits of ECOC from the perspective of the participant organisations?

By answering these questions, the objective of the chapter was to detect case-specific features, as well as more generalisable tendencies concerning the network capital which is mobilised and increased by mega-events.

8.3 Data and Methodology

Network analysis is a method of collecting and analysing data from various individuals and organisations and from their interactions with others. Social network analysis (SNA) as a toolkit is in itself a rather quantitative methodology that most often needs to be complemented with qualitative data collection in order to be able to describe and explain the functions and potentials of the mapped networks. Also, while networks are not rigid forms of relations fixed in time and space, it is their time and space dimensions which are specifically relevant for regional studies. Accordingly, networks need to be understood as being embedded in specific circumstances, geographical or social determinants, and therefore require a longitudinal research approach.

The analyses below are based on data gathered from online survey questionnaires carried out in Hungary (October–November 2014) and Finland (January–February 2015). Respondents were selected on the basis of their participation in the ECOC projects held in Pécs and Turku, based on the published programme booklets and by applying a snowball method. Because of their temporal existence and their high degree of involvement in the projects, the two official management organisations of the featured cultural years, the *Pécs 2010 Management Centre* and the *Turku 2011 Foundation,* were not included in the survey. However, as indicated by many of the respondents, both organisations are present in the networks as 'receiver nodes'.

The languages of the questionnaires were Hungarian, English, Finnish and Swedish. In order to obtain an optimal amount of data for network analysis, the questionnaires needed to be brief, aiming solely at the respondents' cooperations. As anonymity is not possible in this type of analysis, sensitive questions were deliberately avoided. To implement a longitudinal approach, questions were focused on cooperation during the ECOC events, rather than collaborations which existed either before or after the event. The continuity of collaborative linkages was then examined in the light of brief evaluations by the respondent stakeholders, for example, the effect the ECOC event had had on their professional work. The response rates (taking into account only those who verified their participation in the ECOC) were 24 % in the case of the Pécs 2010 project and 29 % related to the Turku 2011 project.

In the analyses, the *nodes* of the networks are various organisations participating in the ECOC projects. The *links* indicate their connections based on a reference by at least one of the actors to the other in a cooperating pair; and assuming mutuality, there are only 'undirected' linkages in the network. For a deeper understanding of the complexities of the networks, four independent *attributes* (one network-intrinsic and three network-external characteristics) are defined for each of the nodes: the degree (number of links to a particular node), the legal status, the field of activity and the geographical location of a participating organisation. When SNA is applied in regional studies, one often faces the dilemma of determining what constitutes a node (an individual or an institution) or what its geographical location or scale is (Comunian 2011b). For the purpose of this analysis, nodes are seen in most of the cases as organisations (except for a few individual artists who are referred to as 'small artistic enterprises' as reflecting their legal status), and their geographical location was defined on the basis of their mailing address (if not otherwise declared by the respondents themselves).

On a technical level, the majority of the information gained from the survey was processed and visualised using the methods of SNA and by applying NetMiner 4 software. Wherever possible, the quantitative results were related to the evaluative answers and free-form comments given by the respondents. Also, in order to be able to assess the functions and potentials of the mapped networks, findings from previous research conducted in the two case-study areas were considered in the analysis.

8.4 Focus on Nodes: Cooperating Organisations and Their Characteristics

Firstly, the analysis aimed to determine the ECOC projects' impact on networks. In more concrete terms, it examined the patterns and intensity (degree) of cooperation during the event year, especially in relation to the different attributes (categories) of the participants, and including their legal status, field of activity and geographical location. The networks visualised included all of the respondents who confirmed their participation in the event year's programme and fully answered the questionnaire. Inclusion in the calculations and visualisation was regardless of their cooperation activity; therefore, the network charts also show 'isolates'

Table 8.1 Summary of basic network characteristics of the Pécs 2010 and the Turku 2011 European capitals of culture (during the event year)

	Pécs 2010 (226 nodes)	Turku 2011 (297 nodes)
Carrier node (one degree nodes)	78 % (176)	73 % (217)
Ordinary node (degree ≥2)	19 % (43)	25 % (75)
Isolate (degree = 0)	3 % (7)	2 % (5)
Total degree (number of links, undirected)	213	320
Mean degree	1.89	2.15
Standard deviation	2.93	3.12
Max. degree	29	26

Source Survey data (2014–2015) processed by NetMiner 4 software

(unconnected nodes), i.e. those participants in the ECOC projects who declared their individual engagement in the events. The size of the cooperation network is somewhat larger in the case of Turku (297 nodes and 320 total links), mainly due to the higher number of respondents, but also due to the more intensive networking activity which is reflected in the higher share of ordinary nodes, i.e. respondents that have two or more links with others. In other aspects, however, the two networks are very similar in their general numeric characteristics (see Table 8.1).

In order to reveal actual patterns regarding the partnerships surrounding the ECOC programmes, the scrutiny of the networks of genuine social-economic actors cannot stop at a mere numeric comparison of carrier, ordinary and isolate nodes, but also needs to proceed further along node dimensions. For this purpose, the attributes of *geographical location*, *legal status* and *field of activity* were also considered. In terms of regional networking (Fig. 8.1) and cooperation between organisations of different legal statuses (Fig. 8.2), the two ECOC projects show more similarities, while the connectivity of participants from different fields of activities (Fig. 8.3) differs between the Pécs and Turku cases. These initial observations and their implications are further elaborated in the more detailed analysis of so-called aggregate networks, where participants of the same class (based on one of the three categories mentioned above) are collected into a single node, and where their links are also combined on that aggregate level.

The regional dimension: The regional networking effect of the Pécs and Turku ECOC projects seems to be very similar. In both cases, the central nodes are the title-holding cities with a high level of internal networking (i.e. bonding ties). This indicates that the performances, exhibitions, etc. within the ECOC projects tend to primarily rely on the local–regional scale. Even though the main slogan of the Pécs 2010 programme was ‘Gateway to the Balkans’, the results show that the engagement of organisations from the West Balkans is not particularly strong. At the same time, the European scale is more represented in terms of participant organisations in the Pécs 2010 project than in the Turku 2011 project. This is a result of the considerable levels of participation and cooperation by organisations of those Hungarian minorities living in the Central European region.

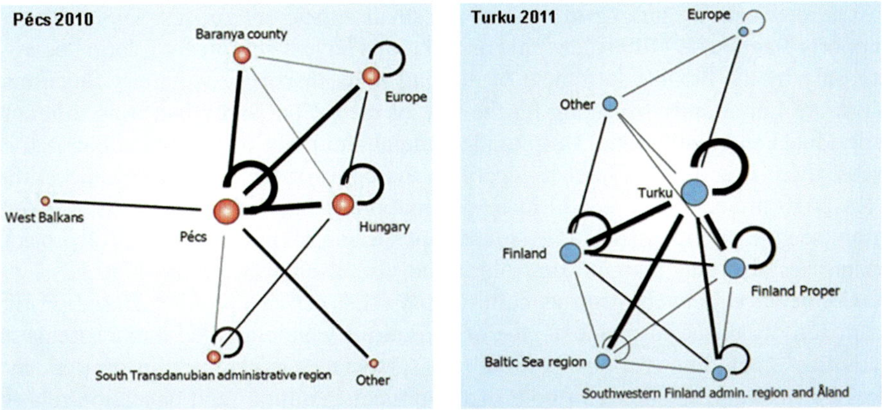

Fig. 8.1 Aggregate networks of the Pécs 2010 and Turku 2011 projects showing the 'regional' dimension. *Note*: the size of nodes indicates the number of links of a node on the aggregated level variety of links (i.e. the larger the node, the more different links it has with other groups), while the *thickness* of the links shows the number of connections on the level of organisations summated under the nodes. *Source*: Survey data (2014–2015) processed and visualised by NetMiner 4 software

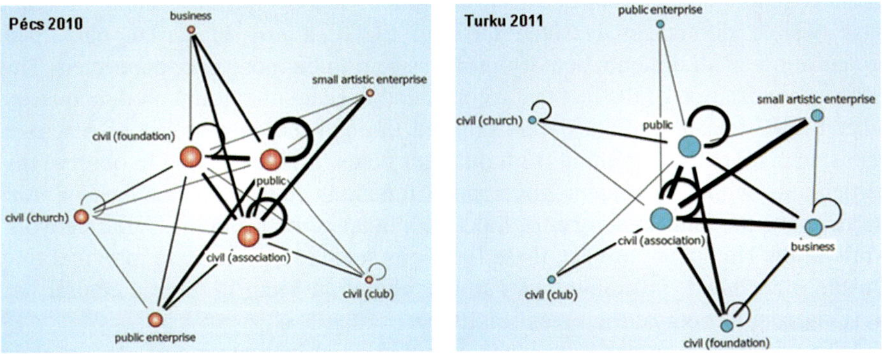

Fig. 8.2 Aggregate networks of the Pécs 2010 and Turku 2011 projects showing the 'legal status' (sectorial) dimension. *Note*: the size of nodes indicates the number of links of a node on the aggregated level variety of links (i.e. the larger the node, the more different links it has with other groups), while the *thickness* of the links shows the number of connections on the level of organisations summated under the nodes. *Source* Survey data (2014–2015) processed and visualised by NetMiner 4 software

Inter-sectorial cooperation: The aggregate networks of the ECOC projects display the constellation of collaborations between public, business, civic organisations (association, foundation, church, club), and public and small artistic enterprises (i.e. individual artists, small bands). In both of the ECOC projects, public organisations (mostly municipalities, publicly funded museums and schools) are in central positions. A diverse range of civic associations are key figures in the

ECOC networks by linking in most of the small artistic enterprises. This evidence supports Bauböck's (1996) idea that associations have a significant role in society, not only by the flexible fulfilment of missing state, market or voluntary functions, but more importantly by taking on the role of mediation, and connecting different individuals and institutions. Despite the general similarity of the two collaborative webs, there is a slight difference regarding the cores of the networks: while in the Pécs 2010 project, there is a clear representation of intensive public-civil cooperation (see the public–associations–foundations triangle), in the Turku 2011 project, businesses also play a significant role in the visualised networking (Fig. 8.2).

Cooperation between various cultural and other activity profiles. Based on the main activity of the respondents, the organisations were classified into 21 fields of activities[3] including various cultural areas, social services, administrative and decision-making functions. In both of the projects, cultural- and education-related activities are the most represented, as well as being the most networked (performing art, visual art, institutions of culture promotion[4]), which is in accord with the main profile of the events. Nevertheless, it is worth looking at the variety of participants with other activity profiles such as local community and urban development, leisure and sports, social services and health care or other services (e.g. financial consulting, retail and professional advocacy) which, although with fewer links, are also part of the collaborative networks which formed around the Turku and Pécs ECOC projects.

As regards collaboration between the various activity profiles, there are some more visible differences between the two ECOC cases. In Turku, participant organisations with different activity profiles seem to be more inter-connected. This can be an indication of the fact that a considerable share of cultural productions and other ECOC-related activities were realised through the cooperation of heterogeneous actors, i.e. those coming from different fields. It is interesting to observe that participants with government–governance functions (e.g. municipalities or ministries) have the highest number of links with other activities in the Turku network, while in the Hungarian project, these functions seem to have a more marginal role. On the other hand, institutions of culture promotion seem to have a central networking role in both of the cities (Fig. 8.3).

When focussing on node attributes, the method of aggregation of organisations on a higher, cumulative attribute-level provides interesting insights into the cooperation patterns of assorted ECOC participants. Beyond some place-specific characteristics such as the higher connectedness of businesses and the more central role of public administration in Turku, and the stronger European dimension of the Pécs project due to the Hungarian quasi-diaspora, there are also several similarities between the two cases: both are highly localised networks with strong local bonds, have intensive public-civil cooperation and show the significant integrative power of civil associations.

[3]The majority of the activity types are present in both cases; 20 types are represented in the Pécs 2010 project and 18 in the Turku 2011 project.

[4]The category of institutions of culture promotion refers to museums, galleries and culture centres.

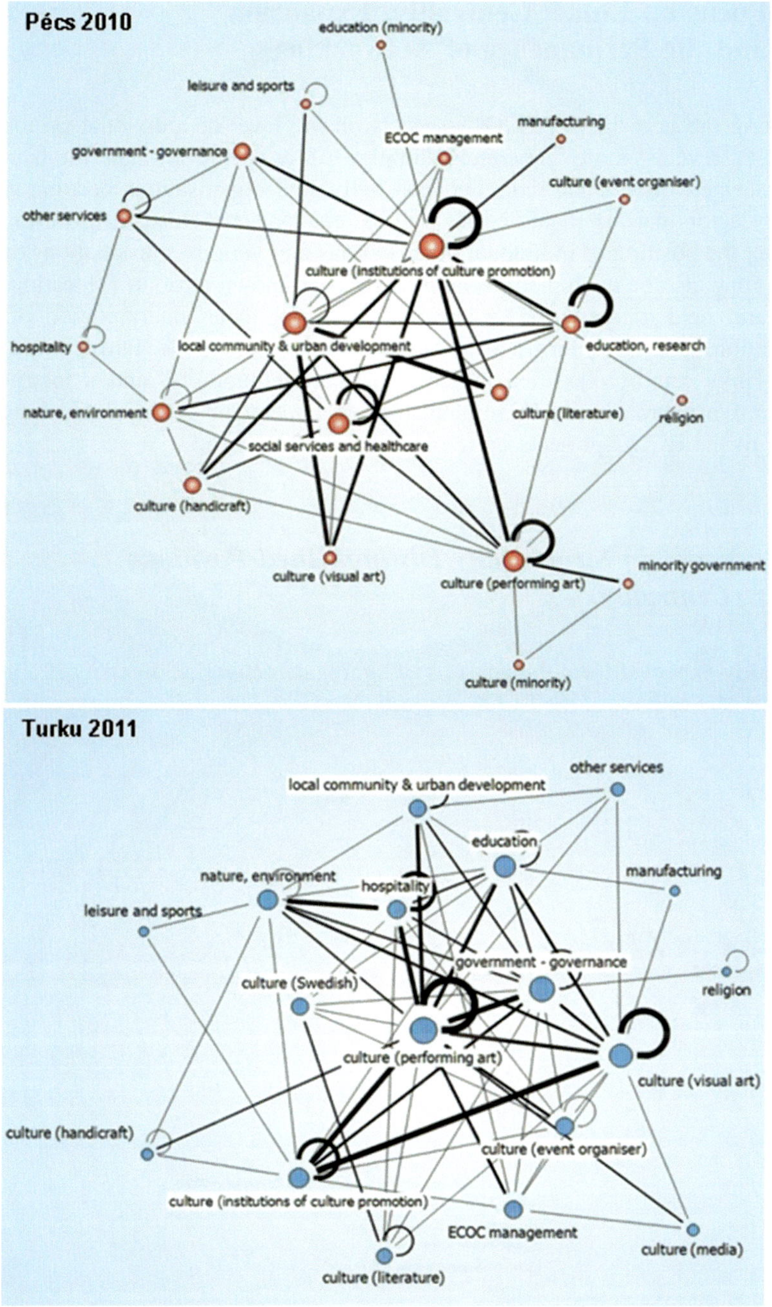

Fig. 8.3 Aggregate networks of the Pécs 2010 and Turku 2011 projects showing the 'field of activity' dimension. *Note*: the size of nodes indicates the number of links of a node on the aggregated level variety of links (i.e. the larger the node, the more different links it has with other groups), while the *thickness* of the links shows the number of connections on the level of organisations summated under the nodes. *Source*: Survey data (2014–2015) processed and visualised by NetMiner 4 software

8.5 Focus on Links: Centrality, Expansion and the Permanence of Networking

Analysing the networks of ECOC projects on the level of individual respondents (Fig. 8.4) reveals more concrete examples of cooperation than the study of aggregate networks, since links between individual organisations become visible, and key actors can be identified as well as smaller actor-clusters. From this perspective, the position of individual organisations and various constellations of actor groups may also be examined in the light of their known attributes (location, legal status and field of activity), so bringing us closer to an interpretation of these cooperation patterns in terms of local development potentials. Furthermore, individual links can be classified in terms of their temporality, and a longitudinal comparative approach can be implemented to understand the potential long-term impact of the ECOC projects on the communities involved.

8.5.1 Nodes in Structurally Distinguished Positions (Centrality)

As already expected from the analysis of aggregate networks, diversity is a general characteristic of both of the ECOC projects. However, there are some specific types

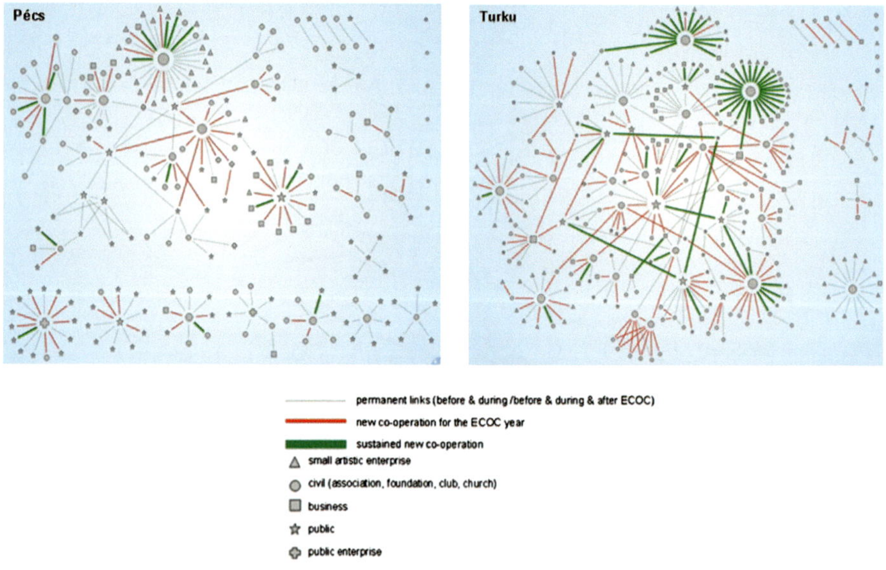

Fig. 8.4 The dynamics of the cooperation networks related to the Pécs 2010 and Turku 2011 projects. *Source*: Survey data (2014–2015) processed and visualised by NetMiner 4 software

of participant organisations that fill particular roles and distinguished positions. Looking at the networks in terms of geographical location/scales, it is clear that the majority of the cooperation clusters are regional mixes, although with a dominance of the title-holding cities as central, high-degree nodes. In the case of the Pécs 2010 network, however, there are also some more visible smaller clusters with purely local cooperation links, with little bridging and more bonding in terms of city-internal and city-external connections. Another aspect concerns the position and distribution of non-cultural organisations within the two networks. In regard to sectorial categories, civil stakeholders (associations, foundations, clubs) have significant roles in both of the networks. In the case of Pécs 2010, almost all the major nodes (and especially those that are interconnected in a wider network of other large nodes) are civic organisations. In the network of the Turku 2011 project, the public sector has a generally more central position, and civic organisations tend to be major nodes and have an important mediator role, i.e. in linking different types of participants and their hubs. Finally, when looking at the classification based on the fields of activity, the share of clusters with an exclusively cultural activity profile is somewhat higher in the case of the Turku ECOC project, but even those cultural clusters are usually connected to larger networks via other kinds of organisations (e.g. from the fields of hospitality, education or public administration/government).

Furthermore, structurally vital positions in the cooperation networks can be filled by organisations whose significance may be less obvious at a first glance. These actor-nodes do not necessarily have the highest degrees (i.e. the greatest numbers of links to others), but may still be the ones that keep the larger structures together, and without which the more extensive networks would fall apart into smaller groups. By calculating the so-called betweenness centrality values[5] of nodes, these cohesive actors can be easily identified. The results of the centrality analysis and the characteristics of the main groups of central nodes are comparable for the two cases and are summarised in Table 8.2. The network of the Pécs 2010 project can be characterised by five main and important cohesive organisations, while the Turku 2011 ECOC project's major central node is the city of Turku (surrounded by other 9 organisations with high betweenness centrality values but still remotely following the node of the Turku city administration). The results of the analysis indicate that these key networking organisations are local civic and public entities, not necessarily from the field of culture but with various activities ranging from education to social services.

[5]Node betweenness centrality quantifies the number of times a node acts as a bridge along the shortest path between two other nodes. The more times a node appears in the paths, the higher centrality it has.

Table 8.2 Network characteristics of the Pécs 2010 and Turku 2011 projects based on centrality analysis

	Pécs 2010	Turku 2011
Number of most significant central* nodes	5	1 (+9)
Regional distribution of the most central* nodes	Very strong dominance of Pécs/Turku city administrations	
Legal status of the most central* nodes	Civil (foundations and associations)	Public (very significant), civil (associations and foundations)
Activity profile of the most central* nodes	ECOC management, institutions of culture promotion, local community and urban development, social services, governance/administration	ECOC management, governance/administration, education and research, nature and environment protection

Main central node characteristics are compared to the rest of the nodes in terms of regional distribution, legal status and activity profile
Source Survey data (2014–2015) processed by NetMiner 4 software
*Calculated based on betweenness centrality

8.5.2 *Expansion of the Networks*

When discussing cooperation networks, their dynamics are probably the most interesting aspect to analyse. This is especially important if one wishes to see the extent to which ECOC projects are able to initiate new and revive former relational spaces through their cultural projects, and whether those relations have the potential to build temporary collaborative structures, which may persist beyond the event years. For this purpose, data are still best analysed on the level of the individual respondents, but the real focus now turns towards the *time frame of relations*. In this respect, three types of relations need to be differentiated; *permanent links*, *new* (*but not continued*) *cooperations* and *sustained new cooperations*. Permanent links that existed either before and during or before-during-after the ECOC event year cannot be considered as indicators of actual network expansion, nor can they be seen as markers of the sustainability of network growth. Thus, only the two latter types of cooperative relations (i.e. networking which was brought about by the ECOC events) are analysed (highlighted in Fig. 8.4).

Besides using already existing networks, the ECOC projects may be assumed to have been significant motivators for new cooperations. In this respect, the Turku 2011 project has proved to be more successful than Pécs 2010 by having significantly more newly created cooperation links. However, it is interesting to take a closer look not only at the number of the newly emergent links, but also their nature and composition. One aspect is the balance between bonding and bridging ties which existed before and were created as a result of the ECOC events (see the above section on 'social networks and regional development'):

- In terms of the participant organisations' legal status, the Pécs 2010 project seems to have a stronger effect (i.e. there is greater added value in terms of the hybridisation of partnerships). In Pécs, the newly established cooperations are mostly of the bridging type, mainly between the public and business sectors or civil groups and small artistic enterprises (while the older permanent links mainly bond actors of the same sector). In contrast, although the ECOC project has brought about new bridging linkages across sectors in Turku, the pre-event partnerships already consist of a balanced mixture of both bridging and bonding types of connections (see Fig. 8.4).
- Regarding the participant organisations' activities, the share of bridging ties is significantly higher in Turku within the ECOC-initiated network, but this kind of diversifying effect is less visible in the case of Pécs 2010.

All in all, it becomes evident that the European Capital of Culture programme had increased the number of bridging links (connecting different types of actors with various resources) in one way or another. Based on Newman and Dale's (2005) concept of the resilience of networks, this may actually have a positive effect on the longer-term sustainability of collaborative networking.

8.5.3 Permanence of Emergent Networking

In addition to its potentials to mobilise cooperative linkages, the prospects for lasting networking capital of the ECOC projects can also be examined. In terms of sustained new cooperations, the Turku 2011 project performs better by having more links between participant organisations that survive beyond the ECOC year (35.7 % as opposed to 20 % in the case of Pécs). It is possible to examine the endurance of these new cooperative linkages in terms of the variety and combinations of different participants they connect; however, it seems that the organisations' legal status does not have an influence on the sustainability of new links (Fig. 8.4). Due to the high number of categories (21), the fields of participant organisations activity cannot be assessed in terms of their impact on the permanence of their relationships.

Finally, the data allow for testing the effect of the geographical location (regional scales) of the participants on the sustainability of collaborative links between them. In this respect, it can be seen that the Turku 2011 project initiated new relations with its wider regions, especially in the Baltic Sea region, and to a much smaller extent in Europe. However, these cooperations seem only to last for the event year and the majority of sustained new cooperation links remain in Turku itself, and to a lesser extent within the Finland Proper[6] (Varsinais-Suomi) region as well as within

[6]Finland Proper is a region made up of 28 municipalities, and its capital and biggest city is Turku. Its Finnish name is Varsinais-Suomi, and in English, it is referred to as Finland Proper. The next larger spatial unit in this study is the state administrative region of Southwest Finland composed of Finland Proper and Satakunta region.

Finland (mainly Helsinki). In comparison, only about half of the new relations initiated by the Pécs 2010 project are local, while the other new cooperation links are from all of the different spatial levels (from county level to those outside Europe). In contrast to the Turku 2011 project, the few sustained relations that are to be seen in Pécs are not restricted to the city, and moreover, their durability is atypical of locally established new relationships.

8.6 Increased Network Capital and Inclusive Governance: Does Networking Matter?

The above analyses confirm that the European Capital of Culture programme has a measurable impact on inter-organisational cooperation and networking among a diversity of actors and that this reaches beyond the field of culture and the limits of the hosting cities. However, the question remains: what do the existing relations which have been mobilised and the new collaborative relations that have been triggered mean to the participants? Also, to what extent is the contribution of an ECOC project perceived to have a positive impact on developing the network capital of organisations and their region? In order to find some direct or indirect relationships between the dynamics of networking and the experienced benefit of the cultural event, the survey contained a request for an appraisal by participant organisations of the effects that the ECOC project had had on their activities. Respondents could select from four options: (a) ECOC hindered their work, (b) ECOC had not affected their work, (c) ECOC meant new opportunities for the event year, but it had no effect in the following period, and (d) ECOC had supported their work in the last 4 years as well.[7] The answers to this question were compared to the respondents' networking activities before, during and after the ECOC year in both Pécs and Turku. The results show (Table 8.3) that the sense of benefitting from these ECOC projects generally increases with the intensity of cooperation during the event year. In other words, the higher the degree of a participant organisation (i.e. the more connected it is), the more positive they assess the impact of the ECOC programme.

Similarly, there is a positive correspondence between an increase in the cooperation links of an organisation (degree growth), and perceived new opportunities. In this regard, there is an observable difference between Pécs and Turku: while in Pécs, growing networks are perceived as beneficial for the organisations' activities mainly during the event, the responses from the Turku 2011 indicate that intensive networking promotes lasting benefits which stem from Cultural Capital projects, i.e. the project has a positive effect on the respondents' activities *beyond* the ECOC year.

[7]None of the respondents selected option (a) in their answers.

Table 8.3 Networking compared to the evaluation of the Pécs 2010 and Turku 2011 ECOC events

Response	Pécs 2010		Turku 2011	
	Average degree during the ECOC year	Average degree growth for the ECOC year	Average degree during the ECOC year	Average degree growth for the ECOC year
(b) No effect	1.1	0.2	2.8	0.8
(c) New opportunities only for the ECOC year	4.2	*2.5*	5.1	*3.1*
(d) Beneficial beyond the ECOC year	5.2	1.8	6.5	*3.8*

Source Survey data (2014–2015) processed by NetMiner 4 software
Italics indicate high average degree growth

Table 8.4 Evaluation of ECOC and the experienced impact on individual organisations' networking

	Experience of impact on an individual organisation's network							
	Pécs				Turku			
(a) Groups of respondents classified by their ECOC-evaluative responses	Isolates (%)	I (%)	II (%)	III (%)	Isolates (%)	I (%)	II (%)	III (%)
(b) No effect on the organisation's work	*44.4*	*44.4*	11.1	0	*33.3*	*33.3*	33.3	*0.0*
(c) New opportunities only for the ECOC year	*33.3*	*38.1*	19	9.5	5.8	5.8	*76.5*	11.8
(d) Beneficial beyond the ECOC year	4.2	41.6	*29.1*	*25*	2.6	15.8	34.2	*47.4*

Source Survey data (2014–2015) processed by NetMiner 4 software
Typology: isolates. no cooperation; I, same cooperation network before and during the ECOC year (and often following it); II, existence of at least one new cooperation link for the ECOC year; III, existence of at least one sustained new cooperation
Italics indicate high percentage

The most interesting results are revealed by looking at the relations between the *sustained* networks (or sustained growth in cooperative partnerships) and the evaluation of the event. For this purpose, nodes have been classified into 'isolates' and types I, II, III based on their highest-value link. For instance, when an organisation has at least one link that represents a 'new (but not continued) cooperation', but no new linkage sustained beyond the event year, it is classified into group II (see more under Table 8.4). This way of processing the network data brings the *experience* of the varying continuities of networking to the surface. The evaluative answers about the events' effects have been compared against the existence of increased networking (or lack of an experience thereof) in order to see whether obtaining new and lasting partnerships is related to the experienced benefits of the ECOC programme (Table 8.4).

The analysis shows that those organisations that have no new or existing cooperations for the ECOC year do not perceive the event as beneficial for their activities. This trend is especially visible in the case of the Turku 2011 project. Additionally, figures indicate that in both of the cities, perceived new opportunities for the ECOC year might be affected by the occurrence of new relations. However, when it comes to the experience of any longer-term positive impacts of the cultural event, the role of sustained relations can be associated with lasting benefits only in the case of the Turku 2011 ECOC (Table 8.4). This may indicate that there is generally a higher significance and value associated with networking (or increasing one's network capital) for actors engaged in the ECOC event held in Turku than those in Pécs.

The generally positive evaluation of the ECOC′s impact on networking is also well reflected in the final optional comments given in the survey. Project-related networking is especially appreciated by organisations with high degrees (5–16 cooperative relations) during the ECOC event, but the possibility of developing expertise in collaborations is also seen as an advantage by those with a lower number of links. The long-term benefits of lasting relationships are directly mentioned in many comments of the Turku 2011 questionnaire and stem from a variety of participants such as the Municipality of Kaarina, Illume Ltd (a company specialised in film production) or the Mixed-Art Association 'Poike'. However, some refer to a lack of willingness on the part of partner organisations, or the lack of resources for maintaining the cooperation. These comments could easily be associated with the inbuilt thematic bias of the survey, i.e. its obvious focus on the relationship between the ECOC programme and networking. In the case of Pécs, under identical survey conditions, there is only one positive comment regarding this relationship (by the Hungary–Japan Friendship Society—Kubaru Group), and this confirms the conclusion made above that participant organisations in the Turku ECOC project tended to perceive more benefits from networking than their Pécs counterparts.

8.7 Conclusions and Policy Implications

There is a richness of tests and measurements offered by the SNA method that can be applied to relational data such as the inter-organisational partnerships that are formed in the context of an ECOC mega-event or other similar events within wider urban governance processes. This chapter presents only a limited number of analyses, illustrating the heterogeneity of collaboration and the dynamics of emergent networks. Whilst there is a need for other sources of data and other analytical approaches, by quantifying and visualising the network of cooperations among diverse ECOC participants, SNA may help to get to the heart of some of the questions related to the general patterns of cooperation networks surrounding mega-events, as well as their locally specific features.

Firstly, the above analysis sheds light on the commonalities and differences in the observable patterns in cooperation networks of the Turku and Pécs ECOC participants. Although the Turku project's collaborative network is slightly more connected than that of Pécs 2010, the levels of networking based on basic network characteristics (i.e. the number of participants and links, and the mean degree) are comparable. The analyses showed in both cases that the involvement of a variety of participants with other than cultural profiles is not only possible but also necessary for connecting cultural actors in the collaborative networks of ECOC projects. These stem from the fields of local community and urban development, leisure and sports, social services, health care or other services, and organisations that could relate their activities to the aims of ECOC. In many cases, these could form heterogeneous partnerships based on resource sharing and complementarities in their roles or objectives. The resulting mix of constellations greatly contributes to the innovativeness and attractiveness of individual ECOC projects.

Furthermore, what has become evident from the longitudinal analysis of this network data is that ECOC projects can contribute to the growth of network capital. As raised in the conceptual introduction on mega-events and social networks, the ECOC programme (which deliberately prefers cooperative projects in its allocation of resources) inherently builds upon and expands collaborative networks. There is evidence that some of these new cooperations have been maintained years after the events in both Turku and Pécs had concluded. These sustained relations are likely to contribute to a better inclusion of social partners (non-public actors) in the decision-making processes in these cities, and also enrich their local–regional governance networks. Thus, an ECOC project can support longer-term local–regional development via the soft 'social' infrastructure of inter-organisational collaborative networks that it triggers.

The sustainability of structures created for and by large-scale events is probably one of the most discussed issues, and it is also relevant in the case of soft, social infrastructures. The ECOC works with and initiates innovative cooperations, but their project's effects on local–regional development largely depend on their durability beyond the actual event year. The analysis of the two case studies indicates that the ECOC programme increased the number of bridging links that connect different types of actors, which is considered as a good general indicator of the future sustainability of networks. Some direct evidence of sustained networks (in the scope of 3–4 years after the ECOC events) is visible, especially in the Finnish cultural capital, Turku. The permanence of cooperations does not appear to depend on the participating organisations' fields of activities, although some regular patterns can be detected in terms of the regional scales of networking: in the manner of outgoing and receding tides, projects spill over during the main event year to cover wider regions and even to cross-national borders, and then withdraw mostly to the confines of the title-holding cities.

However, the comparison of the two cases indicates that mere top-down initiation or external incentives to cooperate are not enough in themselves, and an existing culture of networking is an important contextual factor. The collected data indicate stronger traditions of networking in Turku than in Pécs, and it has a detectable effect on the growth and sustained growth of cooperation links generated by the event. This network-mindedness may be a more general societal phenomenon in Finland (e.g. in the light of the more horizontal social and institutional structures that are present in the country); however, whether there is an underlying wider societal–cultural difference between Finland and Hungary is a question that would need further investigation to answer with any certainty.

As for the experienced benefits of the ECOC projects, results of the analysis show that ECOC-facilitated networking is perceived as a valuable asset by the participating organisations. Their reflections indicate that the recognition of the importance of networking is often associated with the perceived success of the ECOC project, which partly supports the hypothesis that this type of cultural mega-event may strengthen social infrastructures. However, this association was drawn only from the responses of those participating in the Turku 2011 project, which again raises the question of the overall embeddedness such events have in wider societal–cultural settings.

Though the positive local development effects of mega-events are often a feature of political and academic discourses, attention is still mostly paid towards material factors such as the generated profit compared to the size of investment, or the use of created urban infrastructures for increased tourism and use by local residents. However, some of the less tangible impacts are also important. The European Capital of Culture programme's possible contribution to the mobilisation and increase in 'network capital' of individual participant organisations, as well as that of its hosting city and region is an intangible, yet potentially long-lasting positive influence. Considering the fact that the ECOC programme can generate lasting networking relations, as indicated by this research, it can potentially contribute to the strengthening of local–regional governance networks. The extent of this gain varies between ECOC projects and depends on several factors related to the local management of the event, as well as factors which are embedded into wider sociocultural contexts. Therefore, both the ECOC 'practitioners' (on local, national and EU levels) and those who are involved in the impact assessment of mega-events should pay more attention to the potential contributions such events can make to the development of soft infrastructures, and especially the possible enhancement of network capital in the hosting city and its region.

References

Ansell C (2000) The networked polity: regional development in Western Europe. Governance 13 (2):279–291. doi:10.1111/0952-1895.00136

Bauböck R (1996) Social and cultural integration in a civil society. In: Bauböck R, Heller A, Zolberg AR (eds) The challenge of diversity: integration and pluralism in societies of immigration. Aldershot, Avebury, pp 67–132

Bourdieu P (1983/1986) The forms of capital. In: Richardson J (ed) Handbook of theory and research for the sociology of education. Greenwood, New York, p 241–258

Bulkeley H (2005) Reconfiguring environmental governance: towards a politics of scales and networks. Polit Geogr 24(8):875–902. doi:10.1016/j.polgeo.2005.07.002

Comunian R (2011a) Rethinking the creative city: the role of complexity, networks and interactions in the urban creative economy. Urban Stud 48(6):1157–1179. doi:10.1177/0042098010370626

Comunian R (2011b) Social network analysis. Reg Insights 2(2):3

Cropper S, Ebers M, Huxham C, Ring PS (eds) (2008) The Oxford handbook of inter-organizational relations. Oxford University Press, Oxford

Edizel Ö (2014) Mega-events as a catalyst for place-marketing: exploring Izmir EXPO 2015 candidacy and Istanbul ECoC 2010. Reg Mag 295(1):14–15. doi:10.1080/13673882.2014.11006052

Evans GL (2011) Cities of culture and the regeneration game. J Tour Sport Creat Ind 6:5–18

Getz D, Andersson T, Larson M (2007) Festival stakeholder roles: concepts and case studies. Event Manag 10(2):103–122. doi:10.3727/152599507780676689

Gomes P, Librero-Cano A (2014) Evaluating 3 decades of the European Capital of Culture programme: a difference-in-differences approach. http://www.eco.uc3m.es/~pgomes/Papers/Culture_AL_PG_031214.pdf. Accessed 7 Oct 2015

Hall CM (2006) Urban entrepreneurship, corporate interests and sports mega-events: the thin policies of competitiveness within the hard outcomes of neoliberalism. Sociol Rev 54(s2):59–70. doi:10.1111/j.1467-954X.2006.00653.x

Harrison J (2013) Configuring the new ‘regional world’: on being caught between territory and networks. Reg Stud 47(1):55–74. doi:10.1080/00343404.2011.644239

Herrero LC, Sanz JA, Devesa M (2006) The economic impact of cultural events: a case-study of Salamanca 2002, European Capital of Culture. Eur Urban Reg Stud 13(1):41–57. doi:10.1177/0969776406058946

Hirschi C (2010) Strengthening regional cohesion: collaborative networks and sustainable development in Swiss rural areas. Ecol Soc 15(4):16

Isett KR, Mergel IA, LeRoux K, Mischen PA, Rethemeyer RK (2011) Networks in public administration scholarship: understanding where we are and where we need to go. J Public Adm Res Theory 21(suppl 1):i157–i173

Kilduff M, Tsai W (2003) Social networks and organizations. Sage, London

Krätke S (2010) Regional knowledge networks: a network analysis approach to the interlinking of knowledge resources. Eur Urban Reg Stud 17(1):83–97. doi:10.1177/0969776409350794

Mills BM, Rosentraub MS (2013) Hosting mega-events: a guide to the evaluation of development effects in integrated metropolitan regions. Tour Manag 34:238–246. doi:10.1016/j.tourman.2012.03.011

Misener L, Mason DS (2007) Creating community networks: can sporting events offer meaningful sources of social capital? Manag Leis 11(1):39–56. doi:10.1080/13606710500445676

Morellato M, Williams PW (2014) Network stakeholder management as an antecedent to effective mega-event legacies development. Paper presented at the TTRA 2014 international conference: tourism and the new global economy, Brugge, Belgium. http://assets.conferencespot.org/fileserver/file/263379/filename/037.pdf. Accessed 18–20 June 2014

Müller-Seitz G (2012) Leadership in interorganizational networks: a literature review and suggestions for future research. Int J Manag Rev 14(4):428–443. doi:10.1111/j.1468-2370.2011.00324.x

Németh Á (2015) European capitals of culture—digging deeper into the governance of the mega-event. Territ Polit Gov. doi:10.1080/21622671.2014.992804

Németh Á (2010) Mega-events, their sustainability and potential impact on spatial development: the European Capital of Culture. Int J Interdiscip Soc Sci 5(4):265–278

Newman LL, Dale A (2005). Network structure, diversity, and proactive resilience building: a response to Tompkins and Adger. Ecol Soc 10(1):r2. http://www.ecologyandsociety.org/vol10/iss1/resp2/

O'Brien D, Gardiner S (2006). Creating sustainable mega event impacts: Networking and relationship development through pre-event training. Sport Manag Rev 9(1):25–47. doi:10.1016/S1441-3523(06)70018-3

Palmer/Rae Associates (2004) European Cities and Capitals of Culture: Part I. Study Prepared for the European Commission. Palmer-Rae Associates, Brussels

Powell WW (1990) Neither market nor hierarchy: network forms of organization. Res Organ Behav 12:295–336

Provan KG, Milward HB (2001) Do networks really work? A framework for evaluating public-sector organizational networks. Public Adm Rev 61(4):414–423. doi:10.1111/0033-3352.00045

Putnam R (2000) Bowling alone: the collapse and revival of American community. Simon and Schuster, New York

Richards G (2015) Evaluating the European capital of culture that never was: the case of BrabantStad 2018. J Policy Res Tour Leis Events 7(2):118–133

Richards G, Wilson J (2004) The impact of cultural events on city image: Rotterdam, cultural capital of Europe 2001. Urban Stud 41(10):1931–1951. doi:10.1080/0042098042000256323

Steiner L, Frey B, Hotz S (2015) European capitals of culture and life satisfaction. Urban Stud 52 (2):374–394. doi:10.1177/0042098014524609

Woolcock M (2001) The place of social capital in understanding social and economic outcomes. Can J Policy Res 2(1):11–17

Ziakas V, Costa CA (2010) Explicating inter-organizational linkages of a host community's events network. Int J Event Festiv Manag 1(2):132–147. doi:10.1108/17852951011056919

Chapter 9
Urban Regeneration and Local Governance in Italy: Three Emblematic Cases

Anna Trono, Maria Chiara Zerbi and Valentina Castronuovo

Abstract This chapter considers urban renewal policies involving Italian cities over the last 20 years with reference to public and private measures designed to resolve situations of environmental and socio-economic blight in degraded urban areas. Presenting a case-based empirical examination of private initiatives and public policies adopted by local government, the chapter seeks to verify the capacity for management of public funding by local government and to assess the important role played by the private sector in supporting public bodies in the process of urban regeneration and therefore on urban governance. Three case studies are analysed: two medium-sized cities in the South of Italy, Lecce and Taranto, and a large city in the North, Milan. The first two were the recipients, albeit in different periods, of EU funds linked to the "urban" initiative for the renewal of strategic urban areas and urban development. The sharp differences in the results achieved were essentially due to the management capacity of the local administrations and their public and private partners. Taranto was unable to achieve the envisaged urban regeneration, intended to reverse the consequences of poorly conceived industrial and military policies that were responsible for the devastation of the area. This contrasted with the more positive outcome of the urban renewal strategies adopted in Lecce, where public and private actors invested and continue to invest in the rebirth of the old town and the recovery and promotion of the city's cultural heritage. For Milan, the results are still being evaluated. In many cases, the partnership of public and private sectors facilitated the success of the strategies and measures adopted for the renewal of central and semi-central portions of the city (once occupied by industries or by infrastructure that had become obsolete). Over the last few years, the budgets of local bodies, together with their responsibilities in

A. Trono (✉) · V. Castronuovo
University of Salento, Lecce, Italy
e-mail: anna.trono@unisalento.it

V. Castronuovo
e-mail: valentina.castronuovo@gmail.com

M.C. Zerbi
University of Milan, Milan, Italy
e-mail: mariachiara.zerbi@unimi.it

© Springer International Publishing Switzerland 2017
C. Nunes Silva and J. Buček (eds.), *Local Government and Urban Governance in Europe*, The Urban Book Series, DOI 10.1007/978-3-319-43979-2_9

terms of orientation and monitoring of regional transformation processes implemented in concert with private players, have increasingly highlighted the need for public administrations to base their decisions on solid planning of priorities and management skills that are commensurate with the complexity of the measures to be implemented.

Keywords Urban regeneration · Local government · Urban governance · EU urban policy · Lecce · Taranto · Milan · Italy

9.1 Introduction

The theme of urban regeneration and associated policies is particularly relevant in Italy today and the time appears to be ripe for a critical analysis of the numerous experiences that have occurred since the 1990s in a variety of geographical and socio-economic contexts in Italy. Generally speaking, the cities of northern Europe have been ahead of Italy in tackling the problems of economic decline linked to globalisation and the crisis of entire industrial sectors, finding new opportunities for resurgence in economic diversification and the adoption of urban regeneration policies (see cfr, Shurmer-Smith and Burtenshaw 1990; Clark 1994; Wießner 1999; Bonneville 2005; Couch et al. 2011). Despite having been slower to react initially, since the end of the last century public administrations in Italy have benefited from new laws on the supranational, national and regional level. Highly innovative with respect to traditional planning approaches, the new regulations have introduced more streamlined and efficient procedures for administrative measures aimed at regional transformation.

The measures adopted as part of this new season have had mixed results, which are highly dependent on local and urban governance. A number of developments have been seen as follows: the return of urban expansion, especially in the municipalities on the edges of the big cities (construction being the main method of financing municipal budgets); dispersive urban growth (with increased urbanisation of peri-urban areas); an imbalance between the response to collective needs (which ought to be the aim of regional planning) and the accommodation of private interests. In addition, however, actions have been undertaken which need to be carefully evaluated in terms of their future potential and their ability to demonstrate the principles, strategies and methods used in the phases of planning, implementation and management.

Measures designed to improve the quality of life in all the problematic areas of the city constitute the essence of regeneration policies. They appear to contain an intrinsic element of sustainability, arising from the overall reduction of negative impacts (both environmental and economic) that they seek to achieve (Musco 2009).

The variety of local situations and experiences in Italy provides a rich repertory of cases for investigation, by means of which it is possible to reconstruct the evolution of policies adopted over the last 20 years, the different types of measure applied, their effectiveness, and the principles that have guided their use. It is in this framework that the analysis of cases of urban regeneration presented in this chapter should be seen: it seeks to provide not just a critical reflection on what has taken place, but also to suggest, based on the evidence of the results obtained in the various cases, possible guiding principles that can help orient future actions.

This chapter considers urban renewal measures that have benefited from funding from the European Union and/or state and infra-state bodies, of both a public and a private nature. Examples of measures that have enjoyed the support of European funds (allocated, respectively, under the EU's URBAN I and URBAN II initiatives) presented here include the central areas of Lecce and Taranto. The chapter examines both the problems and the ways in which the resources were used to resolve them, providing a critical assessment of the results obtained. In the case of Milan, the third example presented, the focus is on the evolution of the relationship between public and private partners, particularly in the conversion to new uses of central and semi-central urban areas. Special attention is paid to the process of regeneration of an extensive area (in the Porta Nuova district), once used for transport infrastructure, which has since become obsolete as a result of the great transformations that have affected the city. The objective of analysing the experiences of medium and large cities, in the South and the North of Italy, is thus accompanied by the description of the methods used and the planning approaches that they adopted, in order to identify trends and possible perspectives for application elsewhere. In presenting the three examples considered here, the intention is to initiate reflection on the themes of urban regeneration and urban governance. It is hoped that this may lead to a broadening of the research that makes it possible to move from the presentation of exemplary cases to a more detailed comparison between situations that are analogous in terms of their dimensions and the contexts to which they belong. The organisation of this chapter reflects the research phases described above, individually presenting the three case studies, which are framed within their respective geographical and socio-economic contexts, highlighting the lines of intervention and the results obtained. The chapter concludes by observing that the situations investigated can provide the starting point for reflections of a more general character on the principles and methods of intervention which, in accordance with contemporary debate, should guide the regeneration of cities.

9.2 The City of Lecce: Successful Urban Renewal

Maintenance, restoration and recovery are the hallmarks of an ambitious urban renewal project which Lecce, a city at the centre of a fertile and productive peninsula and the capital of a province at the southern end of the Puglia region in south-east Italy, has been conducting for more than two decades.

Lecce is a city with about 95,000 inhabitants (January 2015), having increased by 12.8 % in the last 10 years, and a mainly tertiary-based economy that no longer looks to families alone but is alert to the needs of companies, innovation and research, the latter benefiting from the presence of the University of Salento. Lecce is known for its lively cultural scene, manifested in a myriad of initiatives and events (congresses, seminars, exhibitions and concerts) and in its rich historic, artistic and architectural heritage. The latter is of growing interest to national and international tourists, making the city one of the most interesting destinations in southern Italy.

Its promotion as a tourism destination is recent, dating back to the 1990s, and reflects a desire to invest in culture as a means of compensating for its undeniably remote geography. Lecce has gambled on the recovery of the extensive reserves of archaeological, architectural, historic and artistic heritage located in the Centro Storico, the Old Town. There, every epoch—from the pre-classical era to Roman times, from the Byzantine Empire to the Normans, Swabians and Angevins, up to the long period of domination by the Aragonese and Spanish—has left its more or less evident mark, although it may now be "dispersed" among the structural homogeneity of the baroque city. The city has seen more than 20 years of restoration projects aimed at churches, mansions, monasteries and convents, which had previously been left to the ravages of time, experiencing neglect, blight and abandonment. Today, Lecce has restored honour and splendour to the exquisite buildings which the expert sculptors and masons of the seventeenth century had decorated with fine friezes and other features, adapting the pale local building stone (pietra leccese) to the exuberance of baroque, an important but by no means the only artistic expression of the city.

There have been numerous restoration and urban renewal measures, but the EU's URBAN initiative has definitely played the most important role. It has restored Lecce's image as a city of high artistic value, drawing artists and craftsmen, and it has launched an uninterrupted season of urban renewal. Urban regeneration strategies have also tackled areas adjacent to the Old Town and decaying parts of the suburbs, in which public and private actors have launched many important initiatives to promote the city's cultural heritage (exhibitions, events, conventions and so on).

9.2.1 The URBAN Initiative

The URBAN Programme was developed in the second half of the 1990s (Trono 1996). With a budget of 13,186,000 Euros, it was addressed to the city's historic nucleus of about nine hectares, delimited by the sixteenth century walls and rich in historic and artistic items of great value, but extremely weak in environmental, demographic and socio-economic terms. Indeed, the Old Town was characterised by almost irreversible environmental decay, accompanied by a progressive ageing of the resident population (barely 6000 people, just over 6 % of the city's population), which had declined by about 70 % in the period 1970–1990. Recently the area has seen the arrival of immigrants from outside the EU (belonging to about 40 different ethnicities), who in some parts of the Old Town account for 20 % of the inhabitants.

Other aspects of the area's economic and social problems included a high rate of unemployment, especially among women; a large number of single mothers dependent on welfare; a high rate of illiteracy; a high rate of drug addiction and a large number of minors at risk of deviance. To this may be added the decline of small craft companies; limited urban furnishings, infrastructure and public lighting; and widespread substandard housing. Environmental action was clearly urgently needed, in order to enable lasting development of the Old Town, specifically intervening where the worst indicators of economic and socio-relational problems were concentrated.

The coordinated activation of complementary measures of various kinds (economic, social and environmental) was considered fundamental and was thus given priority, as was the cooperation of social interlocutors involved in urban development processes (public bodies, companies and not-for-profit organisations).

As in many other cities, in Lecce special attention was paid to the wealth of architectural heritage and the great potential in terms of tourism and culture; the presence in the urban fabric of traditional local craft activities such as the restoration of antique furniture and the production of items from papier-mâché, wood, wrought iron and stone, and so on; and its value as a natural outdoor venue for socialising.

The project developed in accordance with a number of fundamental processes: the creation of new economic activities and employment, the improvement of social services and attention to the quality of the environment.

One of the earliest measures was the promotion of economic activities, with particular reference to local forms of production, by supporting craft activities and the consequent creation of "a craft island", organically inserted in the city's cultural and tourism circuit. Subsidies were granted for the refurbishment and modernisation of the artisan's workshops located in the Old Town, with priority given to those craftsmen already operating, or intending to start operating in the area. Help was also provided for the purchase of assistance and consultancy services for craft companies, on an individual or collective basis. The support for local handicraft was completed by the activation of a "school of Salento craft professions" mainly active in the sector of artistic handicraft. Also envisaged were measures to create new micro companies in sectors regarding the enjoyment of artistic heritage, cultural services, guided tours and tourist hospitality.

The Monastero dei Teatini was made available to local small and medium enterprises, providing showroom and commercial spaces and technical assistance for the implementation of initiatives.

Another initiative enabled the promotion and the dissemination of a culture of equal opportunities in order to combat social exclusion and marginalisation and facilitate the economic and social integration of the women resident in the area. Particular attention was paid to unemployed women, women entrepreneurs and women in social services. Specific training and assistance was provided for them in the field of so-called care work, extremely important for civil development and a better quality of life for families. The ex Conservatorio di Sant'Anna was refurbished for this purpose and now hosts a centre supporting equal opportunities for women and disadvantaged persons (Fig. 9.1). To combat marginalisation and long-term unemployment among the inhabitants of the area and facilitate the socio-economic integration of immigrants, a support centre was set-up in the Palazzo Turrisi Palumbo. The improvement of the environment and infrastructure, which included

Fig. 9.1 Conservatorio di Sant'Anna. *Source*: Euromed Agency

the relaying of the road surface with flagstones, the reorganisation of the underground utility networks, the improvement of street lighting and the installation of new urban furnishings, facilitated the establishment of social and economic activities in the Old Town (Ministry of Infrastructure and Transport 2002).

9.2.2 The Post-URBAN Period in the Suburbs of the City

The urban renewal project implemented by the URBAN Programme was extensive and demanding: from the restoration of the façades and monuments to the recovery of the historic and cultural references that create the spirit of the places (craft, commercial and social activities) and the improvement of collective spaces that facilitate social inclusion and the construction of communities. Architectural renewal boosted an urban and socio-economic recovery whose results can be seen today in the streets and alleyways of the Old Town, teeming with diverse craft and commercial activities, which perhaps disorient the inhabitants but stir the curiosity of tourists and draw them to the area.

The success of the measures in the Old Town was (and still is) largely due to the managerial skills of many private players, but also to the activism and initiative shown by the municipal administration, responsible for launching the process of urban renewal and the creation of a system of interregional, extra-regional and international relations, which includes the receipt of funding from the European Union.

Following the revolutionary regeneration produced by URBAN in the Old Town, the measures adopted in the outer areas of the city were more modest, partly due to the more limited public resources. Regional funding (Bando P.I.R. P. Regione Puglia 2006) was used to renew the areas adjacent to the Centro district ("IV° circoscrizione Rudiae – Rione San Pio" and "Via dei Ferrari: Oltre la ferrovia"), while the Contratti di Quartiere (District Contracts) (2004), intended for the social housing areas of 167 A ("Quartiere San Sabino–L'altro Centro"), 167 B and C ("Quartiere Stadio – Verde Centro") made it possible to improve local contexts by tackling processes of urban decay and social marginalisation in the area. Last but not least, the plan known as the Piano Strategico di Area Vasta Lecce 2005–2015, which activates programmes and projects for Integrated Urban and Regional Development, has launched a strategic project (still being implemented) that envisages the regeneration of the landscape and environment in the urban area of the ex Marco Vito quarries and the urban renewal of a stretch of waterfront in the port area of San Cataldo (Trono 2013).

In addition to the urban regeneration programmes the municipal administration has launched projects for multi-level governance, aimed at bringing institutions and economic and social partners into the management of projects with Balkan and Mediterranean countries. Lecce was thus twinned with other European cities, and regional cooperation plans, projects for the integration of immigrants, intercultural youth exchanges, industrial and tourism agreements and protocols of interest to the environment and cultural heritage all followed.

In addition, of particular interest are the more recent projects linked to innovation, research and technology which seek the revitalisation of the city based on a culture of knowledge and services and on the capacity of local companies to use the social and environmental resources in a way that respects the complementarities between the ecosystems of companies and citizens (Rizzi 2013). The Lecce Smart City project focuses on the needs of both citizens and companies, seeking a general improvement of both the quality of information technology and access to it, involving all the regional players (universities, research bodies and entrepreneurs).

Lecce has made great efforts in recent years to reformulate, revalue, redesign and reinvent the city. However, there remain a number of unresolved problems that will need to be faced in the near future, problems common to many other cities in the Italian Mezzogiorno. These include issues of sustainability (transport, waste and socio-spatial imbalances), to which may be added the city's weak capacity for making itself a point of reference for the surrounding area and its limited influence over the rest of the province, whose economy is experiencing severe difficulties. In order to achieve a greater regional balance, it is necessary to strengthen the links between the city and its outlying towns, working with them to develop common services.

9.3 Urban II: The Case of Taranto

9.3.1 Features of the Urban Area of Taranto

With a population of 203,257 people at the end of 2014, the city of Taranto covers an area of approximately 249.86 km^2 and is located on the Gulf that bears its name at the head of the Ionian Sea. The city is spread across three natural peninsulas and an artificial island, the historical heart of the city, set between two natural basins, Mar Grande and Mar Piccolo. Its very favourable geographical location has been the reason for its strategic and military importance over the centuries. The Maritime Military Arsenal, located on the coastal area of Mar Piccolo and inaugurated in 1889, is the cause of marine environmental degradation due to the release of pollutants (such as heavy metals and PCBs) that accumulate on the seabed. It has a staff of about 200 armed forces personnel and 2300 civilian employees specialised in various branches of ship servicing and maintenance.

Since the early 1950s, Taranto has proved to be strategic not just in military terms (strengthened by the construction of a new NATO military base on the shore of Mar Grande, inaugurated in 2004), but also due to its heavy industry. Indeed, with the establishment of the "IV Centro Siderurgico Italsider" (nowadays known as ILVA), it became a centre of excellence in steel production. Inaugurated in 1965, even today it is one of the largest steelworks in Europe (the integrated cycle production unit covers an area of 15 km^2). There is also a refinery owned by Eni S.p.A in the northern part of the city with a primary refining capacity of 120,000 barrels/day, processing most of the oil produced by Eni's Val d'Agri fields (Basilicata, Italy). In the same area, there are many cement factories (which served the early Sixties' growth in the construction sector) and hazardous waste disposal sites.

Despite the presence of large military and industrial sites and their resulting huge production, Taranto is among the areas of the former objective 1, due to its economy being structurally behind the rest of Europe, which is confirmed by the employment and wage data for the period from 2004 to 2014. The industrial sector employs more people than any other sector in the area, but the percentage of people aged 15–64 in work barely reaches 43.3 %, compared with 57 % in Italy as a whole. The overall unemployment rate stands at 14 % (compared with 9.4 % nationally), having peaked in 2014. Unemployment affects above all young people (aged 15–25), reaching 41 %, compared with 31.6 % in Italy as a whole.

The city suffers from decay and socio-economic imbalances associated with the strong influence of industrial companies and military institutions on its economy, which has suffered from their cyclical crises. Signs of urban and social decay include the abandonment of the old town (since the 1970s its inhabitants have been moving to areas of more recent construction), the presence of highly polluting industrial activities close to housing occupied by workers and others, the ongoing occupation by the military of large areas, the high rate of unemployment and, last but not least, the neglect of local cultural heritage and the decay of the old town (Blandino 1974; De Rosa 1979; Giummo 1986). All these factors combined to ensure that Taranto was fully eligible to be one of the target areas of the URBAN Community Initiative Programme. Indeed, the city met many of the programme's conditions: a high level of unemployment, poverty and exclusion, a high level of criminality, a low level of education, a particularly degraded environment and sharp social and economic disparities among neighbourhoods. The environment and the health of residents are deeply compromised (with high mortality rates from respiratory and cardiovascular diseases and tumours of many kinds), and there is a high rate of child mortality. The standardised mortality ratio, in the period between 1995 and 2009, was up to 15 % above average for both genders (Comba et al. 2012).

The URBAN II Operational Programme for the urban regeneration of Taranto immediately revealed a sharp disconnect from reality. Indeed, the planned urban regeneration aimed to modernise a part of the city without taking account of the obsolete industrial and military policies that continue to devastate the area even today.

9.3.2 The URBAN II Operational Programme for Taranto

The programme for Taranto covered an area of 509 ha with 47,400 inhabitants and benefited from an EU contribution of €12.9 million. The main goal was to recreate the ancient relationship between the city and the sea, and, at the same time, with its core, the old town. The areas involved in the programme thus included Taranto's historical nucleus—the "Città Vecchia"—and the neighbourhoods on each side of it, i.e. Borgo (designed by Murat and built in the late nineteenth century) and Porta Napoli/Tamburi (close to the ILVA industrial site). The programme sought to reshape the whole area via the restoration of its historical parts in order to reconnect the old town with the urban area as a whole, emphasising the value of both its historical roots and the settled community, and to strengthen the bonds that establish and fulfil the notion of a city. The general objective was cross-functional regeneration through public measures (Ministero delle Infrastrutture e dei Trasporti, Divisione 5 2009, p. 19). The Managing Authority for the Taranto URBAN II Community Initiative Programme CIP was the Municipal Council of Taranto but, thanks to the agreement, there were many active local actors: Taranto provincial administration, the Archbishop's Curia, the Port Authority, the Chamber of Commerce, the Naval Command for the Ionian Sea and the Strait of Otranto, the Sovrintendenza and the Public Housing Authority (IACP).

The Operational Programme envisaged forty separate measures, 40 % of which consisted of repairs to paving and sidewalks, 22.5 % of improvements to urban furnishings and 15 % were for the purchase of new properties in order to create places for socialisation and the cultural growth of the city. The refurbishment of municipal properties in order to convert them into places of culture accounted for 10 % of the measures, while 7.5 % were in support of project management and advertising and the remaining 5 % were general measures including "training and educational activities" and "de minimis subsidies to tourism, trade and service enterprises" (Ministero delle Infrastrutture e dei Trasporti, Divisione 5 2009, pp. 68–70).

Fig. 9.2 Taranto, Città Vecchia. Environmental degradation. *Source*: Valentina Castronuovo

The definitive text of the programme states that the Municipal Council of Taranto decided to focus on the rehabilitation of the city's aesthetic value as a "tool to make it more socialised and egalitarian, promoting unity and a civic spirit among the citizenry" (Ministero delle Infrastrutture e dei Trasporti, Divisione 5 2009, p. 42). However, it was surely a mistake to see the area of intervention as a happy island, disconnected from the devastation and the environmental degradation that have made the city what it is and continue to damage it today (Fig. 9.2).

9.3.3 Achievements and Expected Results

In a city where the sea has influenced social, economic and productive life for centuries and where the establishment of a military arsenal and the ILVA steelworks generated employment and forged lifestyles, the chance of getting a steady job in an area affected by chronic unemployment became a symbol of aspiration which revolutionized all social values (De Rosa 1979). Trying to turn back Taranto's social, environmental, economic and cultural evolution (or rather involution) by means of a circumscribed operational programme such as URBAN II could only ever be expected to show incomplete results. The period of URBAN's implementation saw a municipal bankruptcy and a political-administrative crisis, which affected the Programme's application. Therefore, following the automatic reduction of funding announced by the European Commission, the Programme had to be reviewed. The contribution of the ERDF (European Regional Development Fund) was reduced by 15. 8 %, as were the co-financing quotas, bringing the total funds available for the programme to €32.95 million. Apart from the project for a property in Via Pitagora (Axis I, Measure 1.1), which was never carried out due to irreparable structural problems, and the partial disengagement from the "Theatre of Innovation" project (Axis I, Measure 1.1) due to a dispute with the contractor, most of the envisaged measures were implemented, or at least partly implemented (Comune di Taranto 2009).

However, although they were completed, the paving and sidewalk repairs and street furnishings could not make up for the failure of long-term plans such as the repopulation of the deserted central areas, the restoration of decaying historical buildings and the improvement of the environment (partly via the renovation of public water supply and sewage networks). It should be noted that the greatest failures concerned the programme's main focus, the historical centre of Taranto, which was supposed to have been "reconnected to the city via the sociological roots of the creation and promotion of the local community" (Comune di Taranto 2009, p. 8). Moreover, the municipal acquisition of two buildings of historic and artistic value, Palazzo Troilo (thirteenth–fourteenth centuries) and Palazzo Carducci-Artemisio (seventeenth century), did not lead to projects of urban regeneration. Palazzo Troilo is currently walled up and abandoned, while Palazzo Carducci-Artemisio has been squatted by local families.

In general, the URBAN II programme was addressed to a specific part of the city, seeking to make it a better, more noble place. However, the project never sought to achieve a definitive break with the economic forces which had subjected the city to such strong negative impacts for so long. Therefore, it did not succeed in producing significant results or in developing a new awareness of the area's true potential. Taranto's experience demonstrates the need for a different approach to the conception and creation of an ecologically sustainable, socially attractive and culturally inclusive urban environment (Elliott 2013).

9.4 Milan: "The City that Rises"

Milan is a city that changes and renews even the way it is perceived: no longer grey but rather an exciting city that prompts one to discover the variety of its new faces. This urban rebirth can be dated to the beginning of the new millennium, although it also bears comparison with the rebirth that followed the second world war, during which the heart of the city, in August 1943, was heavily bombed. The reconstruction led to the recovery and consolidation of some fundamental historic places (such as La Scala, Palazzo Marino, the Galleria Vittorio Emanuele and the Ca' Granda). However, it also saw the unfolding of new architectural episodes that reshaped entire sections of the city.

For more than 20 years after the war, a small group of architects—the 'maestri' of the twentieth century—competed to transform the urban scene. Thanks to the typological innovations they introduced and the imitation they inspired, their high-impact schemes succeeded in creating a hitherto unspoken architectural language: that of Milanese modernity. Towering over the others were those that we might today call the small skyscrapers: the Velasca Tower and the Pirelli building, which represented two of the main sources of inspiration for subsequent projects. Not just the centre, but also the inner and outer suburbs—then growing rapidly—saw important social housing schemes and advanced experiments in the construction of industrial buildings (Berizzi 2015; Irace and Neri 2015).

Following a period of crisis and confusion regarding the role of the architect and the process of urban planning that was supposed to govern the city's development (which began in the 1970s and lasted two decades), the can-do climate of the post-war years now seems to have returned in the new millennium. Among the factors that have combined to bring this about, at the very least we need to consider the transition of the urban economy from manufacturing to advanced services and the consequent availability of a large reservoir of areas vacated by industry or once used for other functions that have gradually become obsolete or have been transferred elsewhere. To this may be added the rise of so-called integrated or concerted town planning, which has made it possible to overcome the paralysis of previous forms of planning, opening the doors to private capital in property investment. The latter factor seems to have been decisive from two opposing points of view: on the one hand it represented an important stimulus for investment in construction, considered potentially profitable, while on the other it obliged the public administration to partially forego its power to impose its own views regarding development.

9.4.1 The Transition from Industrial City to a City of Services

Like most big European capitals, over the last few decades Milan has witnessed the profound transformation of its economy. The most visible feature of this transformation has been the strong growth of those sectors that are tied to the knowledge economy, which is founded essentially on advanced services and high-tech (or highly innovative) companies. The post-industrial transition is a generator of economic and social opportunities that affirm the role of the city as a reservoir of cognitive resources (in terms of types of population, cultural institutions and key activities), and highlight its centrality on the various territorial levels (metropolitan, national and international). The enhancement of urban centrality is often associated however with disruption arising from the substitution of economic activities and populations and the growth of inequality, which put social cohesion at risk (Ranci 2005). Not even Milan is immune to these situations of tension and crisis, although it has been affected less sharply than other European cities. If we observe the situation more closely, we notice a dual change in its economic system: a change in sectoral composition and a change in the size of the enterprises.

Regarding the former, the city has seen the growth in importance of the tertiary sector and, within it, of sectors whose products (both tangible and intangible) have high knowledge content. The concentration in the urban area of these activities has confirmed Milan's position as a node in the global network (Magatti et al. 2005). While it has seen a progressive loss of employment in the manufacturing sector over the last two decades, the city conserves a highly diversified economic structure and maintains strong links with the wider economic system of the metropolitan area and the region as a whole. We are clearly seeing the joint action of two processes of dispersion (of economic activities and specific sectors of the population), together with the concentration of tertiary functions and new populations that have partly replaced those that have left the city (Mazzoleni 2012).

The population of the city of Milan peaked in 1973 (at 1,730,000), since when it has fallen by about a third, reaching 1,256,000 inhabitants at the beginning of the millennium, remaining roughly stable for the next decade. Given the inelastic supply, house prices and rents have acted selectively on the composition of the population, reducing the traditional presence of the middle classes in the central areas. These areas have been heavily affected by the structural changes in the economy, driving competition among the various functions for the use of space and public assets. The main result has been the concentration of activities linked to the knowledge economy, which has had a positive impact on the "centrality" of Milan, transforming it into a powerful attractor of investment, management functions and cultural initiatives, further enhancing supply in the sector. The other side of the coin has been the sharpening of economic and social inequality, made evident by an increase in the incomes of upper medium and high earners, a fall in the incomes of lower medium earners and stagnation for the incomes of lower earners (accounting for about 62 % of taxpayers). The growing socio-economic inequality is accompanied by a worsening of the quality of urban life due to both severe traffic congestion (arising from the intensification of commuter traffic in the metropolitan area) and the social marginalisation and segregation of the city's suburban population, which does not enjoy efficient transport links to the central areas.

The second important change concerns company size. Indeed, medium and large enterprises have been replaced by a myriad of companies of small and super small dimensions. The individualistic nature of many activities in the advanced services sector explains the fragmentation that has occurred and the consequent transformation of entire Milanese districts that had previously hosted industrial activities, now decommissioned or relocated elsewhere. In contrast, the metropolitan area still hosts companies of medium and large dimensions above all in the sector of communications, telecommunications and large-scale retailing.

9.4.2 The Emerging Areas of Fashion and Design

Careful analyses of the distribution of the population and economic activities, focusing on the urban area, have highlighted on the one hand severe inequality between the various areas of the city from the point of view of opportunities and on the other, the existence of growing tensions linked to the reorganisation of the urban economy in order to compete in global markets.

Given their distribution across the territory, the economic and social inequalities can be described on the basis of two gradients: centre/periphery and North/South, the latter deriving from the unequal distribution of infrastructure, thanks to which industrialisation and urbanisation have developed much more intensely in the northern area. The central nucleus of the city, within the ring known as the Cerchia dei Navigli, hosts the historic headquarters of the large universities, the Stock Exchange, the head offices of the major banks and insurance companies, public bodies and the most important cultural institutions. The resident population is on a correspondingly high socio-economic level, being highly educated (48.3 % of the city's graduates live there). There is a greater presence of single-person families than elsewhere, and a lower presence of the elderly and unemployed. These features, albeit less uniformly, also characterise the semi-central areas (in the band between the Cerchia dei Navigli and the Bastioni), beyond which the differences sharpen. The resident population of the suburbs appears to be highly heterogeneous, due to the mixing of long established settlements and those of more recent construction, together with areas of social housing. Here there are greater percentages of elderly persons and those with low levels of education. These demographic characteristics are reinforced by the presence of economic activities that offer few opportunities in terms of qualified employment. The city's outskirts are mainly residential, except for a few locations hosting tertiary activities, which are like islands in the surrounding urban fabric.

In fact it is precisely in some areas that were once considered to be suburban, on either side of the city's external ring road and in some cases even further away from the city, that the innovative elements of greatest interest are starting to be seen today. The strong competition for space that has long characterised the city's historic heart is now also starting to affect areas further away from the centre, especially as a result of property development in decommissioned industrial areas and spaces that have never been built on. The housing tensions driven by the property market—in areas that were traditionally occupied by office and factory workers—have indeed prompted the less well off, especially the young, to move towards less central residential areas, being replaced by people and specialised activities that are more in tune with the city's new economic basis.

These dynamics have paved the way for the concentration in certain districts (Zona Tortona, Ventura and Milano Sud-Est) of activities linked to the creative industries, with the result that Milan is now particularly well positioned above all in the sectors of fashion and design. Indeed, the traditional locations in the city centre associated with these sectors, such as Brera or the so-called "quadrilateral of fashion", have been joined in the last few decades by new areas outside the centre that offer opportunities based on the availability of large and flexible spaces in areas that are still undervalued by the property market.

Some pioneering initiatives undertaken by the cultural and economic elite, starting as early as the late 1980s, have been widely imitated, sparking the revitalisation of various parts of the urban fabric. These are initiatives of a private nature which proliferate spontaneously, there being no overarching design. Moreover, they are limited to the internal renewal of buildings without structural modifications and are not even visible from the street because they remain surrounded by the old perimeter walls of the factories. First and foremost, they exploit the availability of decommissioned industrial buildings, of medium and small dimensions, and they also benefit from favourable planning procedures since they do not require changes in the designated use of the spaces that they are refurbishing.

One of the main limits of these initiatives, which follow a leopard spot pattern and tend to occupy the pores in the urban fabric, is the still-evident difficulty of involving public spaces (roads, squares and green areas) in a positive spiral that qualitatively and irreversibly improves them in favour of the entire community. During the most significant events (the Fuori Salone dominates over all the others), public areas become attractive due to the attention paid to the outdoor space that functions as a general backdrop. However, as soon as the event concludes, the public space goes back to being opaque and amorphous. The phrase "timed centrality" (Bruzzese 2013) efficiently conveys this brief exposure of a location to the spotlights, which, however, is of short duration, the area then returning to the shadows. The incomplete gentrification of these areas, the temporary character of the events that take place there, the inability of the public administration to support and orient the development with suitable policies explain the incompleteness of the transformations.

9.4.3 *Culture as the Catalyst of Development?*

The implementation of regeneration policies by public administrations faces many obstacles. In the course of the 1980s, at the height of the crisis in large-scale manufacturing and heavy industry, the prospects for intervention were limited by a number of factors: the inability of the public administrations to draw up urban planning rules and procedures that could meet the new requirements, the slowness of the decision-making process, the lack of effective entrepreneurship in the property sector, the stagnation of the market in terms of the conversion of buildings to new uses and, last but not least, the limited availability of financial resources. In recent years, with the weight of the other factors now diminished, funding has become the key issue. This has prompted a search for new forms of governance founded not only on partnership with private investors, but also on the ability to mobilise, where possible and in appropriate forms, the citizenry as a whole.

One line of intervention favoured by public administrations has been culture. As is well known, this is based on the assumption that culture can act as a powerful attractor, able to draw the public to the most unlikely locations, and at the same time to contaminate other activities taking place nearby, thereby increasing their dynamism. The history of some of the initiatives undertaken by the Municipality of Milan clearly illustrates this principle and also how the ways in which it is put into practice have changed over the years.

One of the public sector's earliest attempts to recover decommissioned industrial areas concerned the so-called Fabbrica del Vapore (Steam Factory), located in an area long considered marginal, near a large cemetery and the railway yards behind Porta Garibaldi Station. An interesting example of industrial archaeology, the entire complex was acquired by the Municipality of Milan—after having changed hands several times—in the mid-1980s. The aim was to convert it into a Centre for Contemporary Arts, especially for young people, and to provide a space for experimentation with artistic products and advanced technologies. Inaugurated on 21 February 2001 in the period from 2002 to 2008 numerous workshops were set-up inside it, while restoration work that will provide space for parallel activities (exhibitions, shops, restaurants and so on) is still being completed. The revitalisation resulting from the presence of the "Fabbrica del Vapore" has prompted numerous private initiatives that have been set-up in the surrounding area (the most recent in Via Stilicone, where showrooms and art galleries have opened or transferred from elsewhere), generating further instances of renewal that appear to be expanding due to an imitation effect.

With a high degree of contemporary relevance—considering the opening of the Museum of Cultures (Mudec) on 27 March 2015 in connection with Expo 2015—we come to the second case of intervention by the public administration, which concerns an area of strategic importance for the regeneration of the city: the "quadrilateral" of Zona Tortona (Via Solari, Via Bergognone, Via Tortona and Via Stendhal). The area had already been the object of a rapid transformation from a working class area with industrial and craft activities into a place famed for its nightlife and activities linked to fashion, design and culture. Here the Municipality of Milan is the owner of a large plot (about 7 ha) which it purchased in the late 1980s from the Ansaldo company (a large electrical and engineering group that ceased operations in 1989), on which a large-scale project for the creation of a cultural and museum complex is currently being implemented, albeit at an uneven pace. The first significant result here was the installation of the workshops for La Scala—which had previously been housed in locations scattered around the suburbs—in specially allocated sheds occupying an extensive part of the site (about 2 ha), inaugurated on the 20 February 2001. In 2000, the international competition to choose a design for the museum complex concluded with the selection of the proposal by David Chipperfield (in collaboration with the Piuarch studio in Milan). The proposal was held to be most likely to resolve the difficult relationship between old and new while also affirming its own distinctive values. The investment for the implementation was entirely public (more than 60 million Euros), while for the management of the MUDEC a private partner (Società 24 Ore Cultura) came forward; the company will work with the Municipality in the management of all the organisational and administrative aspects. It is hoped that the substantial investment by the Municipality of Milan will help raise the quality of the public spaces, which to date have seen merely episodic and ephemeral improvements, and that it will be able to act in synergy with the numerous private initiatives that have grown up all around it (including Armani, Superstudio Group) in revitalising the entire district from an economic and social point of view.

9.4.4 Up, Up, Up: The Districts that Aim High

For the season of planning by projects, which enabled the insertion of housing, advanced tertiary activities and cultural institutions in marginalised or abandoned parts of the urban fabric, the most challenging test was the big urban redesign initiatives that are responsible for the city's current appearance.

The crisis of the traditional town planning mechanisms, already evident in the 1990s, increasingly led to the use of agreements between public and private actors for the regeneration of areas abandoned by industry or used by transport infrastructure that had become obsolete. The model adopted was the Integrated Intervention Programme (in Italian, PII), which was designed to overcome the rigidity of previous planning approaches, but also to raise the quality of urban renewal. It sought to achieve this by means of a convention between public administrations and private operators founded on the concession of building land and permits in exchange for a commitment to provide public services and infrastructure. Although not always satisfactory from the point of view of the balance of the agreements and the actual results in terms of construction of the "public city", the PIIs represented a significant opportunity for the renewal of the city.

This model accounts for at least thirty major transformations in the urban fabric of Milan, affecting both central and semi-central areas (such as Citylife, the Maciachini centre, the Bodio centre, Porta Vittoria, Pompeo Leoni and Portello South and North) and suburban areas (such as the Politecnico Bovisa, the Bicocca project, Rubattino and Santa Giulia Rogoredo) stretching as far as the first ring of towns around the city (Milano Fiori ad Assago; Ercole Marelli a Sesto S.G.; Fiera di Rho-Pero). Two large transformations that have acquired an undoubted symbolic value are as follows: Citylife, involving the area of the old Fiera di Milano (the trade fair district, which has now relocated to Rho-Pero) and Porta Nuova, which covers a large area north of the centre, which became available after the decommissioning of railway infrastructure (the ex Varesine railway yards) that had become obsolete. In these two measures in particular—which are the most complex in terms of the mix of functions, the relationship with planned and existing transport infrastructure and the relationship with pre-existing public space and private construction—a new architectural language seems to have arisen, whose distinctive features are vertical lines and height, rigorously innovative planning, respect for the environment and the creation of new urban landscapes.

The recent completion of significant portions of the Porta Nuova project, in time for the opening of Expo 2015, has turned the spotlights on this particular case. One of the biggest projects of its kind to be implemented in Milan, it has brought to the fore the idea of creating a Business Centre for the whole city, an idea raised by all the planning bodies that have followed each other in succession since the Second World War but never put into practice. Confirming the role of this central location is the new building nearby that houses the headquarters of Lombardy Regional Administration, completing the range of hi-level functions which are now present there.

The large area affected by the project, which for years was used by provisional structures for recreation and entertainment, is subdivided into three distinct sections, regulated by different plans (Garibaldi Repubblica; Varesine; Isola), separate but integrated. A fertile complexity characterises the entire project, which emerged from the cooperation of a large group of Milanese architects with undisputed international experts and landscape designers in the conception of architectural spaces and the construction of the new urban landscape. The long pedestrianised avenue, the relationship with the park and the attention to the environmental sustainability of the buildings are just some of the key aspects of the project. As well as the formal language, perceived by every observer, attention should be drawn to certain structural aspects that are the result of precise technical and design choices. Fundamental was the decision to use reinforced concrete for all the load-bearing structures of the towers, in accordance with the most advanced methods, as was the selection of special components for the concrete and the building site operations. It has made good use of expert knowledge of materials, expressing the "classic scientific culture of reinforced concrete that is part of the Italian technical tradition" (Migliacci et al. 2013, p. 4). References to the traditional language of Milanese architecture and innovative international perspectives appear to go hand in hand.

9.5 Conclusions

The geographical and economic literature that forms the background of this research has by now clearly described and interpreted the processes that have modified the economic basis of the main European cities over the last few decades. As the investigation conducted shows, Milan is no exception, reflecting trends that are by now familiar, characterised by the concentration of advanced tertiary functions and the replacement of traditional industrial activities (abandoned or transferred elsewhere) with activities linked to the knowledge economy.

While the above-described trends determine the overall context in which the research must act, it is not, however, on them that our attention has focused, but rather on the regional transformations that they have produced and on the role played by the governance of the cities in guiding these changes. From this point of view, in addition to the case of Milan, the two medium-sized urban areas of Lecce and Taranto are also of some interest. Suffering from urban and social decline evident in the decay of large parts of their central areas, these cities are clearly impacted by the generalised processes affecting society as a whole. However, they are also highly conditioned by local economic and social characteristics and dependent on the capacity of local political leaders for planning and governance, and for activating urban revitalisation initiatives that work for the benefit of local communities and ensure opportunities for regional development.

The starting point for the analysis of the regeneration measures adopted in the three cases examined, seen as part of urban governance processes, was the consideration of two fundamental dimensions: the public or private nature of the initiatives and their outcomes for the community as a whole. In the two southern cities, the measures were conducted as part of the EU's Urban initiative (I and II), for which the governance was broad-based but essentially of a public nature. In contrast, in the case of Milan, the investigation looked at a broader range of cases, including both freely undertaken private measures (generally of limited dimensions) and measures conceived and implemented under the aegis of the public administration—or at least "coordinated" by them in terms of the overall planning vision—and implemented by private investors.

The results obtained show that the city of Lecce, after having formulated a programme deemed to be worthy of EU support, was able to correctly manage both the concrete implementation of the planned measures and their associated financial aspects. It launched initiatives aimed at creating new forms of management of urban space, in accordance with an integrated and shared urban and regional development strategy, involving citizens and operators who were able to help guide the future economic, social and environmental quality of the local area, in a framework of effective and efficient implementation of the measures. The city's variegated cultural offer, involving conferences and seminars, itinerant urban workshops and theatre shows, prompted new entrepreneurial forms of urban governance in the cultural sector.

The city of Taranto also correctly interpreted the spirit of URBAN (from which it obtained a significant sum in co-financing) and drew up coherent programmes for its most problematic districts. However, despite apparently sophisticated governance, the programme's implementation phase was not successful, due first to the fall of the municipal administration and subsequently to operational delays and a lack of oversight. There is a clear relationship between competence and good governance—characteristics possessed by the partnership set-up to implement the regeneration programme in the centre of Lecce—and the resulting success of that programme in terms of positive outcomes for the target district and the entire city. In the case of Taranto this virtuous relationship appears never to have been created. The reasons for Taranto's lack of success should also be sought beyond the purely technical and administrative aspects, which were in any case decisive. The long-running uncertainty over the future of Taranto's port and heavy industry generated a context that was so unfavourable as to have an impact, in various ways, on the initiatives undertaken in the old town (it should be noted that one of the districts affected by the programme, Porta Napoli/Tamburi, lies in close proximity to the steelworks of ILVA).

From the strategies proposed by the URBAN programmes, it is possible to derive some innovative suggestions of a general nature that might usefully be employed in regeneration measures, as has actually happened, at least partly. First and foremost among these is an integrated approach, which entails combining in a single project a range of actions that can involve the region's public and private actors. However diversified they may be, the actions need to include not only construction and town planning, but also economic, social and cultural components. A further aspect is the notion of local sustainable development, which requires a model founded on the knowledge and appropriate exploitation of regional resources. It should be pointed out for example that one factor contributing to the success of the Lecce experience was precisely the rediscovery of—and the support for—traditional handicrafts (papier-mâché, marble and so on), which considerably boosted the revitalisation of the central area. Lastly, the participatory approach represents a highly promising strategy for the success of the initiatives, although the cost, commitment and difficulties of implementation often mean it is neglected (at least in Italy), with an involvement that does not reach the level of the citizens.

Assessment of the results obtained in Milan, in terms of the governance that achieved them, appears to be much more problematic, firstly because (unlike the two southern cities) the initiatives took place over a much longer period, and secondly due to their variety in terms of type (of which the analysis presented here can only offer a few examples). However, the empirical evidence found does allow us to make some considerations. Firstly, recognition is due to the spontaneous micro-measures conducted by private players in areas that once hosted industrial activities, giving a new dynamism to districts that had become economically fragile. The main type of transformation that they have produced has been the refurbishment of buildings (above all their interiors) and their subsequent conversion to craft activities. This metamorphosis has been facilitated in administrative terms by the formal continuity in the properties' type of use, which has made it unnecessary to request and wait for authorisation. At certain times during the year, in the areas where they are most concentrated, some of these pioneering initiatives acquire a short-lived centrality. A contributing role here is played by activities linked to the world of fashion and design and their associated events (thematic salons, shows and parades). However, they do not appear to generate direct and permanent benefits of a socio-economic nature for the local population, nor do they have much impact on the public space around them. After each event, the actors leave the scene, which goes back to its grim and empty existence. Similar considerations could be made, with a few modifications, for the initiatives promoted by the public sector linked to "culture", exemplified in this chapter by the Fabbrica del Vapore and the MUDEC. To differing degrees, both have had knock-on effects on other activities of a creative nature, but the expected benefits, in terms of an improvement of what constitutes the landscape of daily life, have not been seen. Pavements, squares, lighting and urban furnishings in public spaces simply do not seem to have been on the agenda of the public administration. Was this a missed opportunity?

The succession over time of the various types of measure examined allows us to trace their evolution. Apart from the initiatives of a strictly private nature, throughout the 1980s the public measures and those guided by the public authorities were centred on the theme of culture, failing to grasp the "green demands" that the citizens were starting to advance: the need for nature and a higher quality of one's living space. In contrast, in the more recent measures the themes of the environment and the landscape have gradually taken centre stage, acquiring a much higher profile in terms of urban planning.

It is perhaps still too early to give a verdict on the large-scale private-sector property developments, represented here by the project of the Garibaldi-Porta Nuova district, currently nearing completion. What is their effective contribution to the construction of the “public city”? What is their response to the demand for housing? There seems to be widespread appreciation of these bold constructions—providing office space, housing and tertiary activities—which have transformed the Milanese skyline, especially the commitment to sustainable construction, the comprehensive conception of town planning, the connections to pre-existing structures in the surrounding area, the public spaces and the large park, seemingly conjured out of nothing. However, leaving aside the opinions of a mainly aesthetic nature that characterised the inaugural phase, serious doubts are being raised regarding the consistency of the new building with the existing and future demand for property. The stock of both luxury housing and offices may struggle to find buyers in the current climate, affected by recession, while the share of social housing, while present in the complex as a whole, is inadequate. It is simply too soon to express any reliable assessment of this imposing development (or indeed of Citylife, also under construction).

It is precisely in this period, when all eyes are seemingly on the big developments (built with no small delay with respect to the moment of their conception), that new demands are beginning to be heard, shifting attention away from the areas affected by dynamics of transformation of varying intensity, towards “peripheral” and static areas, ignored by private investors. Emblematic examples of this type are the districts dominated by public housing, which, apart from some superficial modifications, have been blighted for decades. Conditions have worsened even further in some cases, as a result of the recession, squatting and geopolitical instability, now generating unprecedented flows of migrants.

The idea of urban regeneration, like the idea of reconfiguring ex novo extensive tracts of territory, is making way for more modest approaches based on the idea of refurbishing what is already built, of “timid” restoration (as it has been called by the Renzo Piano group). The focus is now on limited ameliorative intervention, often planned together with the residents, who participate in the process of modernising and managing existing structures. These visions are highly different, but not necessarily exclusive, and might usefully be woven into current urban governance processes and into the urban policies of the future.

References

Berizzi C (2015) Guida all’architettura, Milano. Realizzazioni e progetti dal 1919. DOM publishers, Berlin

Blandino F (1974) La città vecchia di Taranto: il piano per il risanamento e il restauro conservativo. Dedalo libri Editore, Bari

Bonneville M (2005) The ambiguity of urban renewal in France: between continuity and rupture. J Hous Built Environ 20:229–242

Bruzzese A (2013) Centralità a tempo. Industria creativa, trasformazioni urbane e spazio pubblico a Milano. Planum J Urban 27(2):20–25

Clark TN (1994) Urban innovation. Creative strategies for turbulent times. Sage Publications, London

Comba P, Pirastu R, Conti S et al (2012) Ambiente e salute a Taranto: studi epidemiologici e indicazioni di sanità pubblica. E&P, Irvine, pp 305–320
Comune di Taranto (2009) Rapporto finale di esecuzione al 30 giugno 2009 programma d'Iniziativa Comunitaria PIC URBAN II 2000-2006 Città di Taranto. Direzione Risanamento Città Vecchia e Borgo. Comune di Taranto, Taranto
Comune dei Lecce, Ufficio Piano Strategico (2005–2015). Piano strategico di area vasta lecce 2005–2015. http://valutambiente.pbworks.com/f/Lecce+Piano+Strategico+RA.pdf
Couch C, Sykes O, Börstinghaus W (2011) Thirty years of urban regeneration in Britain, Germany and France: the importance of context and path dependency. Prog Plan 75:1–52
De Rosa S (1979) Quasi vent'anni di Italsider a Taranto: alcune implicazioni. Tesi di laurea in Psicologia, Facoltà di Magistero, Università degli studi di Roma, unpublished paper
Elliott M (2013) The 10-tenets for integrated, successful and sustainable marine management. Mar Pollut Bull 74(1):1–5
Giummo LC (1986) Alle Radici dell'abbandono. La città vecchia di Taranto: da realtà rivoluzionaria, a ghetto sottoproletario a città fantasma. Pietro Lacaita Editore, Manduria
Irace F, Neri G (2015) Milano mai vista. Editoriale Domus, Rozzano
Magatti M et al (2005) Milano, nodo della rete globale. Un itinerario di analisi e proposte. Paravia Bruno Mondadori Editore, Milano
Mazzoleni C (2012) La transizione dell'economia urbana verso i servizi avanzati. Il profilo di Milano. Dialoghi internazionali CCIAA di Milano 17:118–141
Migliacci et al (2013). Porta Nuova Garibaldi a Milano. Analisi di un progetto strutturale complesso. Struttural 178, Delettera WP, paper 15, 2013
Ministero delle Infrastrutture e dei Trasporti, Dipartimento per il coordinamento dello sviluppo del territorio, per le politiche del personale e gli affari generali, Direzione Generale Trasformazioni Territoriali (2002) Pic Urban Italia. Rapporto di esecuzione finale, Roma
Ministero delle Infrastrutture e dei Trasporti, Divisione 5 (2009) Urban Taranto, Roma
Musco F (2009) Rigenerazione urbana e sostenibilità. Franco Angeli, Milano
Ranci C (2005) Problemi di coesione sociale a Milano. In: Magatti M et al, pp 265–299
Rizzi F (2013) Smart city, smart community, smart specialization per il management della sostenibilità. Franco Angeli, Milano
Shurmer-Smith L, Burtenshaw D (1990) Urban decay and rejuvenation. In: Pinder D (ed) Western Europe. Challenge and change. Belhaven Pres, London, pp 125–141
Trono A (1996) Lecce, cambiamenti sociali e sviluppo urbano. Conte Ed, Lecce
Trono A (2013) Culture, environment and society. Strengths, weaknesses and challenges for Lecce City. Plurimondi VI(12):41–64
Wießner R (1999) Urban development in East Germany—specific features of urban transformation processes. GeoJournal 49:43–51

Chapter 10
Involving Citizens in the Reuse and Regeneration of Urban Peripheral Spaces

Gerardo Sassano, Antonio Graziadei, Federico Amato and Beniamino Murgante

Abstract The issue of suburb regeneration is highly relevant today. Current urban planning approaches are mainly focused on large and permanent transformations. These solutions often do not take into account the great dynamism and rapid transformations of cities, making any intervention, in most cases, outdated even before its realisation. Nevertheless, these expensive interventions implemented by local government frequently tend to fail the regeneration of these public spaces. This chapter presents the experience of 'Serpentone reload', a workshop based on participatory reactivation of abandoned and underused spaces and buildings in the 'Cocuzzo/Serpentone' neighbourhood of Potenza in Basilicata, Italy. The workshop focussed particularly on the reuse of the 'Ship', an underground building completed in 2010 but never used because, since then, it has been perceived as an extraneous element, being the result of an imposition and not the outcome of shared choices. The aim of this chapter is to propose a methodology that will allow the reactivation of public spaces, the empowerment of communities and the increase in citizens' interest in planning choices, pointing therefore alternative ways for urban governance and for local government urban policies. The results show that the involvement of people in urban planning and in Placemaking activities raises the level of social cohesion, generating both social benefits and quality of public spaces.

G. Sassano · A. Graziadei
Paesaggi Meridiani Association, 24, Via Pienza, Potenza 85100, Italy
e-mail: gerardo@paesaggimeridiani.it

A. Graziadei
e-mail: antonio@paesaggimeridiani.it

F. Amato · B. Murgante (✉)
School of Engineering, University of Basilicata, 10, viale dell' Ateneo Lucano, Potenza 85100, Italy
e-mail: beniamino.murgante@unibas.it

F. Amato
e-mail: federico.amato@unibas.it

© Springer International Publishing Switzerland 2017

C. Nunes Silva and J. Buček (eds.), *Local Government and Urban Governance in Europe*, The Urban Book Series, DOI 10.1007/978-3-319-43979-2_10

Keywords Public space · Placemaking · Tactical urbanism · Temporary design · Public participation · Urban planning · Municipality · Local government · Urban governance · Italy

10.1 Introduction

Traditional urban planning approaches tend to privilege large-scale transformations of the urban space, which are usually perceived as permanent and final solutions. These solutions often do not take into account the dynamism and rapid transformation of cities, making most of these planning actions outdated even before their realisation. Therefore, traditional urban planning approaches are no longer able to really take into account the great complexity of urban organisms.

La Cecla (2015) states that the real problem of urbanism is that in numerous cases, it is not a discipline of observation, listening and interpretation of urban reality. The representation of complexity is still a set of hatches, flows, zoning where it is difficult to recognise not only a genius locus, but also, and above all, a relationship of belonging and mutual influence between the city and its inhabitants.

In the last century, with the progressive separation of several approaches of urbanism, mainly based on procedural aspects, to people and the material and immaterial relations between them, the role of the planner as an expert able to deal with their 'knowledge' of the complexity of the city has emerged. The decrease in attention to human aspects, the places of everyday life and a vision of the city as a product of its own inhabitants began to be evident already in the text of the Charter of Athens (International Congress of Modern Architecture CIAM 1933). This document highlights the rationalist approach of giving to each person their rightful place and identifying and separating their functions (e.g. housing, work, recreation and traffic movement) through the practice of zoning. The issues highlighted by the Charter of Athens created a divergent vision of urban planning, overcoming the prevailing dimension of the previous century. Mumford (1954), who perceives the city in a more complex dimension as the place where human coexistence is played, extends the humanistic dimension introduced by Geddes (1915). Mumford, referring to the culture of the city, provides an anthropological approach, where the city is considered as one of the oldest products of human activity far from the idea of top-down planning. In the same position, against planning and urban renewal methods adopted in that period in the US, Jacobs' (1969) analysis of American cities is mainly addressed to 'simple and common issues' and to the real-life of the cities. Gehl (2012) extended this approach. Drawing on his theoretical and professional experiences, he highlighted the importance of the activities in urban spaces and social relations.

In Italy, the results of traditional urban development policies produced, in several cases, unfinished or abandoned interventions, unused or useless infrastructures, and public spaces not recognised by the community. Urban transformations are often huge interventions produced with substantial public investment, involving the destruction of large areas of soil and the devastation of peri-urban and rural landscapes and were often carried out without involving the community. In many cases, these heavy interventions did not produce significant or durable improvements in the quality of life of the directly affected local community. These problems have been increasing lately because of the current economic crisis and the associated political choices aimed at a gradual reduction of public services and urban welfare.

The cultural, economic and social conditions that made up city life in the past have been deliberately broken down by the current political and economic context (Berdini 2014). The traditional network of urban solidarity has disappeared, as urban services to people have been drastically reduced, removing the vision of solidarity that was behind urban life in Europe. The financial situation of most Italian municipalities is dramatic and is gradually worsening. Whilst in 2009 the municipalities in official economical default were two and went up to eight 1 year later, by July 2014 this number had grown to 180 (Fubini 2014). None of these municipalities is able to invest resources in the community's needs or in programming new public actions or services to citizens.

One possible answer to this complex scenario is to give importance to communities. On the other hand, the principle of subsidiarity and public participation defined by Article N°118 of the Italian Constitution is the basis of many bottom-up initiatives promoted by local associations and committees. These can thus play an important role in saving the city and reconstructing its material and immaterial skeleton. Therefore, it is necessary for inhabitants to rethink the city, suggesting interventions that are the result of participatory processes and not decisions imposed by few stakeholders.

For citizens, the conquest of participatory spaces requires, in exchange, a great effort in terms of responsibility and civic sense. The way forward is to imagine and build together a new kind of city in which the citizens' engagement would be at the centre of the urban policy.

The aim of this chapter is to propose a methodology, based on a case study, that would allow the reactivation of public spaces, the empowerment of communities, the increase in the citizens' interest in planning choices and the rise of the decision-makers' awareness of the importance of citizens' associations in urban life and in urban governance. The second section of the chapter discusses how community involvement in urban planning processes can be a successful strategy in developing urban regeneration experiences. Section three will focus on the formalisation of the concept of placemaking, defined as a self-building activity of public spaces carried out by the citizens and aimed at raising the level of social inclusion as well as the quality of urban space. The fourth section will introduce the case study, focusing on the complex urban development of the Cocuzzo neighbourhood, in the municipality of Potenza (Southern Italy). Section five will describe the participatory approach and the placemaking activities conducted in the neighbourhood to reactivate its abandoned public spaces. Finally, section six will discuss the how the city can benefit from the physical space and for the communities that the reactivation of public space activities carried out with a bottom-up approach can provide.

10.2 New Strategies for Urban Regeneration

Nowadays, cities need to adopt a strategy based on patching urban mosaic, on regenerating urban spaces and on the re-appropriation of urban places. The main feature of this regeneration path is the adoption of different approaches to planning and urban design, considering the development of social relations as a top priority, with the aim of reimagining both the city and the role of the citizens in it. Urban development and regeneration strategies are therefore strictly related to the social dimension. In the so-called city of individuals, people's biographies, with their social relationships intertwined with the urban space, are relevant and can guide policymaking and administration. The main direction that now arises is to design cities having people as a reference, in order to deal with wishes as well as with needs. In practice, beyond the city itself it is important to consider also the 'look and feel'—in other words, the image and sentiment of a city. When approaching the government of an urban space, it is fundamental to capture its new complementary dimensions, direct expression of the wishes of its inhabitants: actual city, lived city, imagined city and dreamed city.

Citizens' requests for their cities, today, are increasingly related not only to the demand for new services, adequate functions or spreading of urban quality, but also to the claim of everything that can help in making the city a unique and irreproducible heritage. There is the need to produce urban projects based on the growing demand for beauty and attractiveness, which are able to reflect the expectations of the inhabitants. The action will be focussed on urban public spaces, real-life scene, meetings and social interaction. The physical dimension of a city is important, but equally important is its density of social relationships. These are the places where cohesion and inclusive processes occur, the places of cultural norms regulating behaviour, the places of identity expressed through urban public space.

Cities are a system based on complex relationships. They are the place of policy and practice of change and transformation, based on visible (buildings and spaces) institutional and social aspects. The urban political process should be seen as a result of social interactions. Facing and designing urban policy means, therefore, taking into account social problem solving issues, in an effort to find rational and useful solutions that correlate with its inhabitants, city users and spaces.

Reinterpreting urban policy in this dimension means contributing to the reduction of inequalities, the management of expressed or latent social conflicts and the building of a sense of community where it is missing or greatly weakened. Urban planning should be centred on these reflections, through a massive cognitive and participatory mobilisation (Lorusso et al. 2014) able to meet a new right to the city (Harvey 2008), based on the demand of beauty, variety, usability, security, sustainability and so on. The need of an individual and collective welfare is the new paradigm, which is complex to satisfy because it is based more on immaterial aspects than on the physical changes of the city. On the other hand, the same right to the city is not confined to individual liberty to access urban resources but has the right to change us by changing the city. It is a collective right, rather than individual, which inevitably requires the exercise of a common power.

10.3 Placemaking Formalisation

The concept of the city as a common space is widely recognised. Today the involvement of more or less formal groups of citizens is the central issue of numerous interventions on urban space. Among these place-based and people-oriented actions, the experience of placemaking plays a primary role. The term placemaking defines the processes of a citizen's involvement in developing new visions of public spaces. These processes include not only the regeneration of abandoned or underused public spaces, but also to the development of relations among those citizens who take part in the planning process, building new social relations and strengthening their sense of belonging to a community (Chase et al. 2008). The Faro Convention on the Value of Cultural Heritage for Society (2005) recommended these participatory approaches.

In Italy a General Direction for the Contemporary Art and Architecture and for the Urban Suburbs was founded in 2014 as a branch of MiBACT (Italian Ministry of Cultural Heritage and Cultural Activities and Tourism), to improve the architectural and urban quality and the regeneration of urban suburbs. This institution produces innovative approaches in designing these places, no longer conceived as marginal areas, but as essential parts of the cities.

Urban suburbs are places where daily life produces and sediments tangible and intangible values that make urban landscape meaningful. One of the main problems of suburbs is the absence of traditional public spaces. While the city centre is generally characterised by the presence of strongly identifiable public spaces, in peripheral areas these are less common, often designed on highly experimental ideas or by filling empty residual spaces generated by the not well-planned development of residential areas. In the suburbs, disconnections in social structure correspond to physical disconnections of urban space. Today, considering the complex economic situation, it is not possible to overcome the lack of social quality in public spaces of suburbs through the construction of new areas. The only possible answer to address the needs of the community is the development of bottom-up processes using the opportunity offered by pre-existing hidden spaces. The right approach is regenerating such spaces through reduced interventions concerning physical size and development time. This means promoting self-construction operations of space, aimed at defining temporary uses of sites.

The strong relationship between Placemaking processes and social structure creates barriers in formalising a robust methodology as the theoretical foundation for the development of this type of activities (Shibley et al. 2003). Despite this, it is possible to identify some common features in all Placemaking processes. These activities are successful only if the community is self-involved in the process, through voluntary action, or if it generates the Placemaking action itself. The achievement of a citizen's awareness is fulfilled when they feel that the city's space really belongs to them. This cannot be the result of top-down processes, where a policy maker or a highly structured group of people imposes a vision of a city and public space to citizens. To successfully implement these experiences, it is important to involve groups of individuals with different social and cultural backgrounds in the processes. A process can be truly inclusive only when it is not restricted to specific social or economic groups and associations. Particular emphasis has to be given to the inclusion of less represented groups in a society (e.g. women, immigrants and children). If in traditional participation processes one of the stronger risks was the possibility that little more than a simple consultation/information exchange came out from the final analysis, Placemaking ensures that this restriction is overcome, because involved groups directly work on the construction of urban spaces, becoming designers themselves of the physical reality of the city. An important aspect is also the need for processes to have a certain temporal continuity, which does not finish after short activities of space (self) construction, but they are defined by complex paths with many activities aimed at community empowerment. In fact, the main purpose of this type of activity is the impact on social space more than the transformation of physical space (Lydon et al. 2012). These activities have been developed with the aim to improve safety and inclusion of urban communities.

10.4 The Case Study

A placemaking experience has been developed in Potenza Municipality (Basilicata, Italy). In 1998, the Italian Ministry of Public Works published a call, with the aim of low quality suburbs regeneration. The main purpose was to improve the quality of neighbourhoods characterised by a lack of environmental quality and services, paying particular attention to the salvage of buildings constructed in modern periods and energy efficiency improvements (Pontrandolfi et al. 2000). The Ministry selected and funded 46 projects. The project submitted by Potenza Municipality, concerning the regeneration of two neighbourhoods, 'Poggio Tre Galli' and 'Cocuzzo', has been included within this funding (Murgante 2005; Murgante and Danese 2011).

The first neighbourhood is the typical example of a dormitory suburb. The second one is characterised by high complexity in terms of housing with strong social conflicts. This neighbourhood has the largest concentration of housing, despite considerable accessibility problems and an enormous lack of public spaces (Las Casas et al. 1996, 2005). This building, entirely devoted to social housing, is five hundred meters long and forty meters tall and has a sinuous shape, which generated the name 'Serpentone', literally big snake (Fig. 10.1). This huge building based on 'Unité d'habitation' principles became a symbol of urban decay in a suburban area of the city.

Fig. 10.1 The 'Serpentone': **a** A picture shortly after its completion when it was disconnected from the central part of Potenza (*left*), **b** the original project (*right-up*), **c** a recent picture which highlights how urban development has included the neighbourhood in the urban part of the city (*right-down*) (*Source* Municipality of Potenza)

The regeneration project, funded by the ministry in 1998, was based on the following elements:

- lowering of two floors of the building, realising a multifunctional roof with a heating system;
- accessibility improvement;
- realisation of underground parking; and
- creation of a huge square.

In the adjacent area of Poggio Tre Galli neighbourhood, the regeneration project had foreseen the construction of new social housing, a parking area and a neighbourhood square. The whole project was financed by more than 10 million euros by the Ministry, 2 million euros by the local organisation of Social Housing management (Azienda Territoriale per l'Edilizia Residenziale, ATER), 625,000 euros by Potenza Municipality and 4.8 million euros by private investors. However once funding was obtained, the project was radically changed.

Enric Miralles was commissioned to work on the design of a large public space in Cocuzzo neighbourhood. Unfortunately, after several months the architect was diagnosed with cancer, which caused his untimely death. The project was then assigned to the architect Marco Casamonti. His idea of improving the urban space of Cocuzzo neighbourhood was based on the transformation of Via Tirreno, the neighbourhood's only road, into a large park. The proposed project considered the strengthening of a road surrounding Serpentone and Serpentino, a smaller sized building similar to the first located a few meters away from it. In this way, the road that separates the two huge buildings could be destined to be a park (Murgante 2005, 2007).

The main idea was to transform Via Tirreno into a green corridor 'animated by the presence of rocks' in the lower part of the neighbourhood and a 'ship' with huge concrete sails ten meters tall in the upper part. To date, the ship is the only part of the regeneration project realised (Fig. 10.2). This impressive concrete object, since its creation, is at the centre of a strong controversy.

Today, the ship made with significant public investments is not used. The roof garden is totally abandoned, forgotten both by residents and by the municipality, which stopped all kinds of maintenance. The huge 'cargo', the space under the roof garden, was never completed. The ship, therefore, caused a decline in quality of life for residents, who, as a protest against the municipality administration, unable to understand their real needs and requirements, have continued their daily life in perfect indifference, as if the ship never existed. Inhabitants proudly declare that they have never stepped on the roof garden of the Ship.

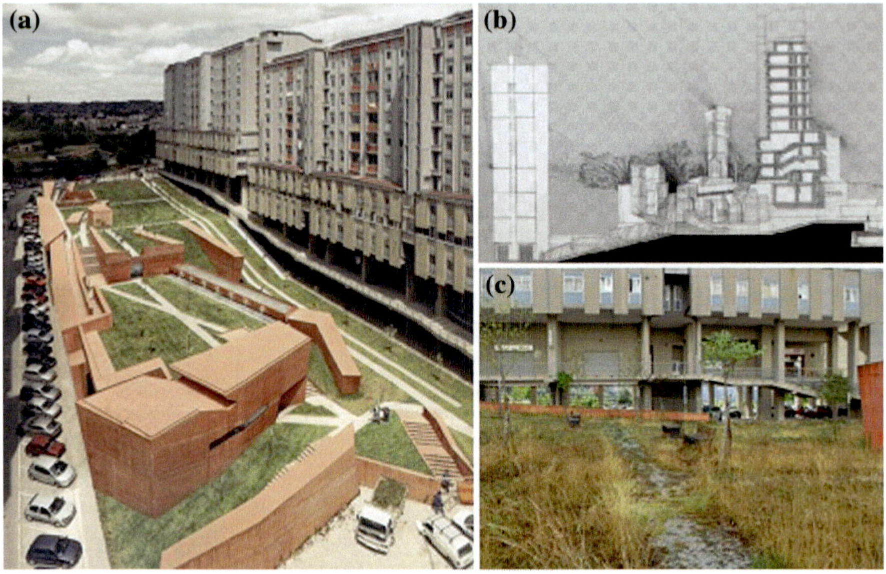

Fig. 10.2 **a** The realised project (*left*), **b** the original project (*right-up*), **c** the roof garden totally abandoned (*right-down*) (*Source* www.archea.it; authors)

10.5 A Placemaking Experience

'Serpentone Reload' experience started with a workshop held in September 2014 organised by the Cultural Association Basilicata 1799, with the aim of studying the 'Cocuzzo' neighbourhood in Potenza (Basilicata, Italy), based on previously described principles. The initiative was aimed to work on the reactivation and regeneration of underused or abandoned sites and stimulate the local community to re-appropriate and maintain it. Specifically, during the workshop, the decision was made to intervene with a Placemaking action on a symbolic place in the neighbourhood, the so-called 'ship'. The choice was not random because the site is characterised by a strong aversion by its citizens, who do not identify themselves with the place and reproach the location and architectural choices. Moreover, a key theme of the experience was to explore the urban landscape and then deal with the perceptive dimension of the places and their communities. This responds to the requirements of the European Landscape Convention (2000): 'Landscape means an area, as perceived by people, whose character is the result of the action and interaction of natural and/or human factors' (http://www.coe.int/en/web/conventions/full-list/-/conventions/rms/0900001680080621). Other intentions behind the initiative were to investigate the relationship between communities, suburbs and cultural heritage.

The initiative aimed to involve people in an active process of interaction and exchange, starting from the assumption that no one better than the inhabitants who spend their daily lives in a place can know it. Only direct knowledge offered by inhabitants can provide valuable information on place dynamics, problems and potentialities. From this perspective, the inhabitants were considered not as simple users of the intervention, but the real experts of the place, the key partners in the process to consult in every phase of the work. Starting from the experience of people, the attempt was to understand problems and potentialities of the place, collectively reimagining the public spaces reinventing and transforming them into the heart of the community.

To emphasise the role played by the inhabitants, the workshop started with a walk in the neighbourhood, conducted by a group of residents and aimed at knowing the places and acquiring information, stories and testimonials (Fig. 10.3). In the following days, they collected views of privileged witnesses who provided their vision of the neighbourhood and their life experiences. These stories revealed that the community held a deep aversion to the project of the 'ship', seen as an imposed external object, and a sense of shame that comes from being resident in a neighbourhood that in the collective imagination of people is perceived as degraded.

Another crucial moment was the meeting with students. Through a playful experience, they expressed a different point of view, free from prejudice and schemes that usually influence the perceptions of adults. Students were encouraged to express, through the game of 'it can', all that was possible to do in the neighbourhood. The game produced interesting scenarios for activities actually carried

Fig. 10.3 **a** Workshop participants guide students from school to the 'ship', **b** few games built by workshop participants were enough to bring the neighbourhood children on the 'ship', **c** a small boat on the roof garden of the 'ship', **d** children were encouraged to express, through the game of 'it can', all that was possible to do in the neighbourhood, **e** the Solarium, space created during the workshop, attracts children and local residents, **f** The ship cargo has become a site of painting exhibition (*Source* Salvatore Laurenzana, photographer, 14–21 September 2014)

out during the workshop and in terms of place perception and its unexpressed potentialities.

After this initial phase, a strategy has been developed in order to encourage the inhabitants to re-appropriate the space. The intention was to absolutely avoid the production of a 'design'. The attempt was to create a 'place' working on immaterial or 'almost nothing', so that people feel comfortable and accepted.

The playful and irreverent approach was a key to play on the metaphor of the 'ship', imagine building a port as a place of relationships and exchanges. So people were involved in the construction of an object familiar to everyone, a paper boat, which opposed its ephemeral lightness to the weight of solid concrete.

The entire community was so involved in the construction of thousands of these objects and their production has become a collective ritual, which for 2 days has been the pretext for the community to get itself together and imagine that the 'ship' could really dematerialise and sail away. Subsequently all the community imagined different uses of the 'ship', usually used as a deposit, and its roof garden. As a result, several playgrounds and play areas have been realised through self-construction, attracting children and allowing an almost completely unused area to be lived in an intensively and shared way. The covered part of the 'ship' has been imagined by the community as a place for social and cultural activities for the whole city. At the end of the workshop in this place, a temporary public space was opened, significantly called N. Av. E. (New Expressive Adventures) which, playing again with the name given by the inhabitants (in Italian nave means ship), and it has allowed the realisation of the dream of seeing the 'ship' sail away towards the territories of creativity. The closing day coincided with the inauguration of N. Av. E., with an exhibition of photographs and videos taken in the neighbourhood. In the following months, the space was used for theatrical representations, shows, exhibitions and conferences, evidencing how, through a participatory and shared path with the community, it is possible to imagine a new life for abandoned places, considered in the popular imagination a 'waste' or 'something unwanted'.

10.6 Social Benefits of Citizens' Participation

The process triggered by the workshop produced changes not only in the use of places, but also in the perception of the place by the local community. From a sense of shame for the inhabitants, a sort of pride in belonging to this place has been produced. They were able to imagine themselves differently, naturally present in the community, and the workshop had only a catalyst function as a cornerstone of Placemaking processes. One of the eleven principles of creating great community places (Project for Public Spaces 2012) is to have a vision (step five http://www.pps.org/reference/11steps/). Here, Sariputra, form is emptiness and emptiness is precisely form (Heart Sutra). This Buddhist quote is a good metaphor for the image of the implementation idea. For Buddhism, vacuum means the absence of itself, not only from a spatial point of view, but also from a time point of view. This implies that each element and phenomenon of the internal or external reality is not only interconnected with the other, but is also provisional and interconnected with the provisional nature of the other elements and phenomena. It determines a structure that is relative, relational and, at the same time, transitory, impermanent. In that 'empty' context, the idea of a space that should not have to tend to a fix structure, but maintains its performative fluidity was experimented.

Continuing with the Buddhist metaphor, the empty quarry, 'Serpentone Reload' was based on the principle of 'almost nothing' or, as it became more common in urban planning, 'Lighter, Quicker and Cheaper' (Bravo et al. 2013; Maynard 2013; Silberberg et al. 2013; Rube 2013; Sassano and Graziadei 2015). All the interventions performed during the workshop were completed with little resources and in few days.

In conclusion, the experience of Serpentone Reload can be representative of how fruitful it is to operate in critical social contexts, producing real changes in places and the community. These transformations can take place in a very short space of time and without the investments that are generally associated with traditional urban interventions.

The experience highlights that people who live in the city must shape it and interventions must be the result of participatory processes, containing bottom-up instances, avoiding decisions imposed by few people. Shared decisions inevitably have greater strength and a better chance of success. From the citizens' point of view, the achievement of participation spaces requires a great effort of responsibility and civic sense towards what is shared. The right approach is imagining and building together a city of space reuse, involvement and sharing, a community place for everyone as the result of a common effort.

10.7 Conclusion

This chapter shows the central role played by urban regeneration in city development policies, and more generally in urban governance processes, particularly in the peripheral areas characterised by the presence of underused or abandoned public spaces. At the same time, in these areas, there is a lack of social cohesion and a deficiency in the inclusion of citizens within decision-making processes. The chapter proposed a Placemaking methodology as a means of creating interconnection amongst social groups, enabling the inhabitants of a place to work directly on the public spaces of their own city. This allows them to have a direct relation with their specific needs and to increase their sense of belonging to a community.

The case study illustrated the experience of 'Serpentone Reload'. The workshop was an innovative experience for different players:

- for decision-makers, because it offered the opportunity to experiment with inclusive policy processes and demonstrated the power of regeneration through minor actions, relatively quick and involving few resources;
- for participants, young designers who have directly verified the possibilities to design a project that actually has not built any artefact, nothing tangible, measurable or quantifiable; and

- for inhabitants, who accepted to put their initial refusal and rejection of the object under study and to trust themselves and their landscapes.

Serpentone Reload called the designers to reflect on their responsibilities and offered the opportunity to rethink the importance of inclusive planning processes, which are often undervalued.

Participation is too often confused with consensus. Decision-makers and designers frequently abdicate their duty of developing sustainable, inclusive, resilient and safe cities. They assume the role of mediators, collecting and nourishing the wish list of their voters or customers. Experiences such as Serpentone Reload show that participation means sharing skills and points of view, finding solutions that are not final but that continually change and grow, taking place through time, modelling on the living body of the city. The involvement of people in the design process of urban spaces is a theme already widely discussed in literature (Nevens et al. 2013). This chapter has highlighted the active participation of citizens in the processes of self-construction of public spaces as a further step towards the creation of communities with a higher level of social inclusion. These Placemaking activities allow not only raising the level of self-reliance of citizens, but also creating a greater sense of belonging to a community. This approach results in a quality improvement of the citizens' social life. Moreover, communities that are more cohesive are generally more interested in the processes of urban governance and urban planning, helping to stimulate decision-makers and administrators to work on improving the quality of the urban space.

Acknowledgments Authors are grateful to Sara Bellarosa, Giuseppe Biscaglia, Luca Catalano, Annalisa Metta, Maria Livia Olivetti, Pasquale Passannante and Francesco Scaringi for the organisation of the activities, to Salvatore Laurenzana for the photographs and to Mimmo Nardozza for the workshop organisational coordination. Many thanks go to all the workshop participants: Sara Adesso, Giorgia Botonico, Bernardo Bruno, Antonio Canosa, Simone 'ilovetu' Cortese, Giuseppe D'Emilio, Simona Di Giovanni, Francesco Fazio, Michela Galletti, Serena Lorusso, Mariangela Meliante, Rocco Morrone, Carmen Peluso, Daniela Policriti, Lydia Postiglione, Mariangela Russo, Alberta Caterina Santagata, Chiara Terranova and Elenonora Vaccaro.

References

Berdini P (2014) Le città fallite. I grandi comuni italiani e la crisi del welfare urbano. Donzelli, Rome

Bravo L, Carmagnini C, Matityhou N (2013) Lighter, Quicker, cheaper: towards an Urban Activism Manifesto. In: XXVIII Congeresso Nazionale Istituto Nazionale di Urbanisitica, Salerno

Charter of Athens (1933) IV international congress for modern architecture (CIAM)

Chase JL, Crawford M, Kaliski J (2008) Everyday urbanism. Monacelli Press, New York

European Landscape Convention (2000) http://www.coe.int/en/web/conventions/full-list/-/conventions/rms/0900001680080621. Accessed April 2016

Faro Convention on the Value of Cultural Heritage for Society (2005) https://rm.coe.int/CoERMPublicCommonSearchServices/DisplayDCTMContent?documentId=0900001680083746. Accessed April 2016

Fubini F (2014) I Comuni italiani verso la bancarotta: 180 a rischio default. La Repubblica, 21 luglio 2014

Geddes P (1915) Cities in evolution: an introduction to the town planning movement and the study of civics. HardPress Publishing, Lenox

Gehl J (2012) Vita in città. Spazio urbano e relazioni sociali. Santarcangelo di Ro-magna, Maggioli

Harvey D (2008) The right to the city. New Left Rev 53, Sept–Oct 2008

Jacobs J (1969) Vita e morte delle grandi città. Saggio sulle metropoli americane. Einaudi, Torino

La Cecla F (2015) Contro l'urbanistica. La cultura delle città. Einaudi, Torino

Las Casas G, Argento R, Pontrandolfi P (1996) Un'esperienza di urbanistica partecipata-comunicativa lungo il fosso di S. Lucia a Potenza: i primi passi del laboratorio urbano. In: Contributo dell'INU al Rapporto italiano per la Conferenza ONU Habitat II. Istanbul, giugno 1996, Presidenza Consiglio dei Ministri

Las Casas G, Pontrandolfi P, Murgante B (2005) La riqualificazione del Vallone di S. Lucia a Potenza: l'efficacia di programmi ed interventi per il recupero di un'area marginale della città. atti della XII Conferenza internazionale "Vivere e Camminare in Città. Le periferie" Bergamo—Brescia, 9–10 giugno 2005

Lorusso S, Scioscia M, Sassano G, Graziadei A, Passannante P, Bellarosa S, Scaringi F, Murgante B (2014) Involving citizens in public space regeneration: the experience of "garden in motion". In: Lecture notes in computer science, vol 8580, pp 723–737

Lydon M, Bartman D, Garcia T, Preston R, Woudstra R (2012) Tactical Urbanism, short term action||Long term change, 2. Retrieved on 3 Oct 2013 form www.issuu.com/streetplanscollaborative/docs. Accessed 3 Oct 2013

Maynard M (2013) Light, quick and cheap: the big shift in urban planning. Forbes. http://www.forbes.com/sites/michelinemaynard/2013/10/23/light-quick-and-cheap-the-big-shift-in-urban-planning/. Accessed April 2016

Mumford L (1954) La cultura delle città. Edizioni di comunità, Milano

Murgante B, Danese M (2011) Urban versus rural: the decrease of agricultural areas and the development of urban zones analyzed with spatial statistics. Int J Agric Environ Inf Syst 2 (2):16–28

Murgante B (2005) Le vicende urbanistiche di Potenza. EditricErmes, Potenza

Murgante B (2007) La barca di via Tirreno va: Dal contratto di quartiere alle polemiche di questi giorni. Il Quotidiano della Basilicata

Nevens F, Frantzeskaki N, Gorissen L, Loorbach D (2013) Urban transition labs: co-creating transformative action for sustainable cities. J Clean Prod 50:111–122

Pontrandolfi P, De Fino C, Guida A, Guida F (2000) La riqualificazione possibile. Periferia, metodo, progetto. Il Contratto di Quartiere della zona occidentale di Potenza, Ermes Edizioni, Potenza, pp 4–368

Project for Public Spaces (2012) Eleven principles for creating great community places, http://www.pps.org/reference/11steps/. Accessed March 2016

Rube K (2013) Lighter, quicker, cheaper remaking streets for better outcomes. Project for Public Spaces 2013. http://www.pps.org/. Accessed March 2013

Sassano G, Graziadei A (2015) Cambiare il volto della città a partire da piccole (e concrete) azioni. Il Quotidiano della Basilicata 2015. http://www.ilquotidianodellabasilicata.it/news/cronache/733877/Cambiare-il-volto-della-citta-.html. Accessed

Shibley R, Schneekloth LH, Hovey B (2003) Placemaking in the binational niagaras. J Architect Edu, 57(1): 28–42. doi:10.1162/104648803322336566

Silberberg S, Lorah K, Disbrow R, Muessig A (2013) Places in the making: how place-making builds places and communities. Department of Urban Studies and Planning (DUSP) MIT. http://dusp.mit.edu/cdd/project/placemaking. Accessed April 2016

Part III
Citizen Participation in Local Self-Government

Chapter 11
Assessment of Socio-Economic Status Relevance for Latvian Electoral Participation

Lilita Seimuskane, Inga Vilka and Edgars Brekis

Abstract In many countries, there is increasing concern about the decrease in the level of electoral participation in all types of elections. The situation in Latvia is not an exception. Despite the fact that the first scientific studies of electoral participation analysis date back to the 1940s–1950s, it is still relevant today to understand and explain the factors that affect the citizen's choice (not) to participate in elections. The goal of this chapter was to find the factors that influence participation in elections in Latvia, and indirectly how it may affect local government and urban governance processes in the country. Using data from two elections in Latvia—local elections in 2013 and elections to the European Parliament in 2014—the authors examined whether there is coherence between factors of participation intensity and socio-economic status (education, income, employment, and so on). In total, data of 22 socio-economic variables were tested, which are used by the government for planning and organising the development policy of local governments. To achieve the goal, the authors used a logistic regression approach. Analysis showed that level of education is a significant factor, both in the European Parliament and in the local elections. But socio-economic factors correlate more clearly with participation in the European Parliament elections, namely income of the resident's birth rate and region of residence. Participation in local elections is affected by contextual factors that are specific to each local government.

Keywords Local government · Local election · European Parliament election · Socio-economic status · Voter turnout · Citizens' participation · Urban governance · Latvia

L. Seimuskane (✉) · I. Vilka · E. Brekis
University of Latvia, Riga, Latvia
e-mail: lilitase@inbox.lv

© Springer International Publishing Switzerland 2017

C. Nunes Silva and J. Buček (eds.), *Local Government and Urban Governance in Europe*, The Urban Book Series, DOI 10.1007/978-3-319-43979-2_11

11.1 Introduction

Electoral participation is one of the oldest and most popular types of political involvement of citizens. Despite the fact that the first scientific studies of electoral participation date back to the 1940s–1950s, it is still relevant today to understand and explain the factors that affect the citizen's choice (not) to participate in elections. Many researchers (e.g. Duch 1998; Bendor et al. 2003; Feddersen 2004; Geys 2006; Stoker 2006; Pacek et al. 2009; Smith 2009) refer the shrinking confidence in institutions of democratic representation, distrust in politicians and political parties as the main reasons for the diminishing level of electoral participation, and low voter turnout in elections. The Latvian situation is not an exception. Voter turnout continues to decrease in all types of elections, including to the Parliament, local elections, and to the European Parliament.

According to the results of the public opinion survey performed by Eurobarometer, the population in Latvia tend to trust in local governments more than in the Parliament or in the national government. After examining the results of three autumn surveys by Eurobarometer, it can be established that on average 47 % of the population trust in local governments, 17 % trust in Parliament, and 22 % trust in the Government (Table 11.1). Besides, the indicator for trust in local governments in all three study periods is higher than the average in the EU Member States, thus differing from the indicator for trust in the Parliament and Government (Table 11.1).

Although trust of the residents in local government is higher, when compared to trust in the Parliament and Government in Latvia, the voter turnout in the local elections when compared to the one in the parliamentary elections is lower (Fig. 11.1). Voter turnout in the local elections of 2013 was the lowest since the restoration of Latvia as an independent and democratic state. Average voter turnout was 46 %. In 20 local government units, the turnout reached 50 %, but the voting level in less active municipalities was below 33 % (Vilka and Brekis 2013).

In reply to the question about the most effective ways to protect their interests in local government and their institutions, most of the respondents in Latvia stated elections (Seimuskane and SKDS 2012). However, voter turnout is reducing in all types of elections, including local (Fig. 11.1); at the same time, one can recognise the growing trust to all three types of public power after 2012 in Latvia (Table 11.1).

Table 11.1 Citizens' trust in public authorities in Latvia (%)

	2012		2013		2014	
	Tend to trust		Tend to trust		Tend to trust	
	Latvia	EU27	Latvia	EU28	Latvia	EU28
Regional or local public authorities	45	43	48	44	49	43
National parliament	13	28	17	25	22	30
Government	17	27	21	23	28	29

Source European Commission Standard Eurobarometer No. 78 (2012); No. 80 (2013); No. 82 (2014) data

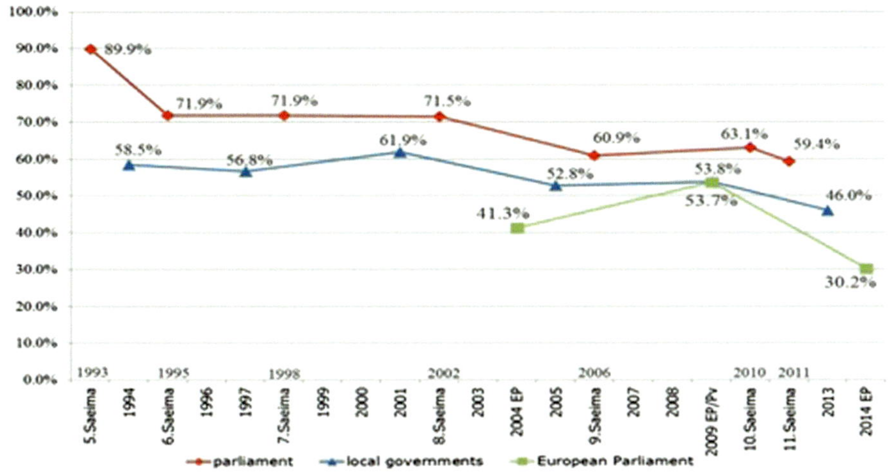

Fig. 11.1 Voter turnout in elections to the Parliament, European Parliament, and local governments in Latvia (1994, 2004, 2010). *Source* The Central Election Commission of Latvia 2015

Although studies of citizen participation in local elections have a long tradition (e.g. Pateman 1970; Sartori 1970; Brady 1999; Van Deth 2001), it is relatively recent in Latvia. To fill this gap is the aim of this study. The goal of this chapter was to uncover the factors that influence electoral participation in Latvia and through that on local government and in urban governance. Using data from two elections in Latvia—local elections of 2013 and elections to the European Parliament in 2014—this chapter examines whether there is coherence between factors of participation intensity and socio-economic status (education, income, employment, and so on). In total, data of 22 socio-economic variables were tested through logistic regression. This chapter contains five sections, including this introduction. Section 11.2 contains the theoretical aspects of political participation. Section 11.3 describes the results of elections in Latvia used for analysis as well as the background basic information about local governments and the territorial division of Latvia. The results of the analysis are included in Sect. 11.4. Section 11.5 is the conclusion.

11.2 Factors Affecting Local Election Turnout

The social structure of society determines to a large extent the patterns of political participation (Almond and Verba 1989; Verba et al. 1995; Rosenstone and Hansen 2003; Hooghe and Quintelier 2013). Inequality based on class position, education level, income, age, or gender has a strong impact on the level of electoral participation. This relationship between the social structure and political participation can also have strong democratic consequences, as it implies that some groups within societies will be more successful in getting their voice heard in the political decision-making process than other groups (Hooghe and Quintelier 2013). Verba and his colleagues obtained three answers when studying why individuals do not participate in politics: they cannot, they do not want, and they have not been asked to participate (Verba et al. 1995). Assessment of the three groups of answers allowed the above researchers to identify the factors that could affect participation of the individuals: appropriate resources, motivation, and mobilisation (Verba et al. 1995). They called these subjective factors of political participation Civic Voluntarism Model. According to this model: "(..) individuals are more likely to take part when they can, when they want to, and when they are asked (..)" to participate (Schlozman 2002:439). Still this does not mean that political participation is determined by all three factors. However, this indicates that in case there are no motivation and mobilisation, the presence of resources alone will not automatically facilitate political participation (Schlozman 2002).

The response, *they cannot*, characterises insufficiency of resources—time and civic skills to take part in participation activities, money to contribute to campaigns and other political causes (Verba et al. 1995). The key principle of the socio-economic model is that the individuals with a higher socio-economic status, those who have a higher level of education and income, as well as a job of a higher status, are more active participants (Verba et al. 1995).

The various empirical studies in Europe confirm that the effect of the socio-economic status is still stable as for the forms of political participation. The Citizens' Audit in Britain (2004) and European Social Survey (2006/2007) conducted in 25 European societies confirmed that political engagement is very much dominated by the already well-resourced, the most highly educated, the rich, and those from the top educational echelons. It shows clearly that for all types of political activity, those with higher levels of education are more likely to engage and get involved: those reaching the tertiary stage reach consistently higher levels of engagement than those whose schooling was finished with primary or secondary school (Stoker 2006; Keil and Gabriel 2013). A number of leading researchers approve also the effect of income level on the population participation. Belgian researchers, Marien et al. (2010), indicate that one of the latest participation forms —political consumerism—in its direct form depends on availability of financial resources that grants freedom to an individual to choose the goods regardless of their price. The impact of family income on political participation has been broadly analysed also in studies of Verba et al. (1995). The higher the income, the more the intensive is political participation.

Although of all the factors that affect political participation socio-economic resources have been examined in a more detailed way in different countries, still the Civic Voluntarism Model by Verba et al. (1995) describes the subjective status of political participation of each individual formed by their resources of the individual level. Contextual factors—size of municipality, affiliation to a particular administrative territory with its specific conditions, and customs—are just as important.

In a very recent study on size and local democracy (Denters et al. 2014), the authors mention three groups of claims about the democratic effect of population size—one that stresses the negative effect of the increasing population size, another one that reflects the positive effect of the increasing population size, and the third line of argument that implies that population size is not likely to have any effect—either positive or negative—on the quality of local democracy. This possibility arises because population size may covary with the predominant socio-demographic composition of individuals living in different municipalities (Denters et al. 2014).

11.3 Local Elections 2013 and European Parliament Elections in 2014: A Comparison

11.3.1 Background Information on Local Governments and Territorial Division in Latvia

The 2009 administrative-territorial reform in Latvia replaced the two-level local government system with more than five hundred units, by one-level local government system with 119 units—9 cities and 110 municipalities (*novads*) (Fig. 11.2). In this reform, 92 units resulted from the merger of the previous local government units, while 27 units correspond to the previous units that did not change (Vilka and Brekis 2013). A republican city is characterised by a well-developed commercial activity, transport facilities and public utilities, social infrastructure, significant cultural institutions, and at least 25 thousand residents. Municipalities differ in terms of the area, number of population, density, economic features, and level of development. According to the law, each municipality should have at least 4000 permanent residents, but in fact more than one-third of municipalities have less population than that. Rīga is the capital of the Republic of Latvia and contains almost one-third of the population of Latvia. The local elections in 2013 were the first elections in Latvia after the first 4 years of work of the reformed local government system.

Besides that, there are five planning regions in Latvia—Rīga, Vidzeme, Kurzeme, Zemgale, and Latgale, which act as coordination and cooperation institutions among the municipalities and between the municipalities and national institutions. The Rīga planning region was divided into two statistical regions due to its size and heterogeneity: the statistical region of Rīga, which includes the city of Rīga, and Pierīga, including the rest of the territory of the Rīga planning region. Therefore, there are six statistical regions: Rīga, Pierīga, Vidzeme, Kurzeme, Zemgale, and Latgale (Fig. 11.3). These statistical regions are the smaller territorial units for the collection and analysis of statistical data in compliance with the Nomenclature of Territorial Units for Statistics of the European Union (State Regional Development Agency 2011).[1]

[1]In this study, we will use these statistical divisions.

Fig. 11.2 The administrative-territorial division and local governments in Latvia. *Source* Vilka and Seimuskane (2012)

Fig. 11.3 The territories of statistical regions in Latvia (*Source* Picture from publication of Latvian Academy of Science, Vilka 2013)

11.3.2 Local Elections of 2013

In Latvia, a local government council is elected for a period of 4 years in equal, direct, secret, and proportional elections. The number of deputies to be elected to the council is determined in accordance with the number of inhabitants registered in the administrative territory of the respective local government on the election day. Persons having the right to vote, who are citizens of Latvia and of the European Union Member States, having reached the age of 18, are entitled to participate in the local elections. In order to participate in the local elections in Latvia, a citizen of the EU should be registered in the Population Register of Latvia. Persons who are serving sentences in places of deprivation of liberty, as well as persons who do not have the right to vote in the EU Member State whose citizens they are, have no right to participate in the local elections.

On 1 June 2013, the sixth local election was held in Latvia since the restoration of independence. A total of 46 % voters participated therein. 8725 deputy candidates were registered for the elections, and 1618 deputies were elected to the municipal councils. The highest voter turnout was in Rīga—56 % followed by Vidzeme—45 %, Latgale—42 %, Kurzeme—41 %, and Zemgale—40 %. The local governments with the highest voter turnout, as well as their location according to the breakdown by statistical regions, are listed in Table 11.2.

The highest voter turnout in the local elections in 2013 was observed in Mērsrags Municipality (Kurzeme region)—62 % of those who had the right to vote participated in the elections, which was a significantly higher voter turnout when compared to the overall turnout in the state constituting 46 % (a boxplot, Fig. 11.4).

This was due to a particular development of events in this region. In 2008, i.e. a year before completing the reform, Roja and Mērsrags parishes (*pagasti*) voluntarily merged into a single administrative-territorial unit—Roja Municipality. Creation of Roja Municipality was facilitated by opportunities provided by the Administrative-Territorial Reform Law to local governments of Latvia concerning a

Table 11.2 The Local governments with the highest voter turnout at the local elections in 2013

No.	Local governments	Voter turnout (%)	Region
1	Mērsrags Municipality	62	Kurzeme
2	Rugāji Municipality	60	Latgale
3	Vaiņode Municipality	57	Kurzeme
4	Ērgļi Municipality	56	Vidzeme
5	Pārgauja Municipality	56	Vidzeme
6	City of Rīga	56	Rīga
7	Rucava Municipality	56	Kurzeme
8	Vārkava Municipality	56	Latgale
9	Alsunga Municipality	55	Latgale
10	Aglona Municipality	54	Latgale

Source Summary made by the authors using data of the Central Election Commission of Latvia

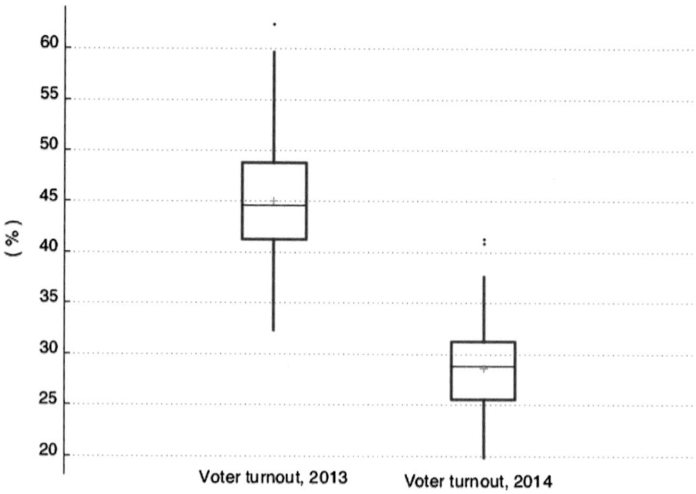

Fig. 11.4 Distribution of municipalities by voter turnout in the local elections of 2013 and elections to the European Parliament in 2014. *Source* A boxplot developed by the author using data of the CEC of Latvia on voter turnout

single grant from the national budget amounting to 1–5 % of the total amount of annual budgets of amalgamating municipalities when they merged voluntarily within the term set by law. Unfortunately, the idea of merging local governments did not result in cooperation in solving daily problems among the deputies of the former parishes. A part of the residents of the former Mērsrags parish considered that their interests in the newly formed municipality were not taken into account and signed a request to their deputies for the previous parish to secede from Roja Municipality. In autumn 2010, the Parliament of the Republic of Latvia (RL) adopted the law On Division of Roja Municipality and Launching of Operation of the Newly Established Municipalities. In December 2010, new elections were held in the already divided municipalities.

The development of events regarding the merge of parishes and organisation of complex procedures resulting in separation of the local government facilitated dynamic action and mobilisation of residents of Mērsrags Municipality to participate in the local elections. These local governments' territories are not dropped as outlying data, since increased activity of citizens in elections is determined by internal political and socio-economic factors.

11.3.3 European Parliament Elections in 2014

The European Parliament is elected once in 5 years in direct elections. In 2014, the third elections of EP deputies were held since Latvia joined the European Union

Table 11.3 Local governments with the highest voter turnout in the elections to the European Parliament in 2014

No.	Local governments	Voter turnout (%)	Regions
1	Ikšķile Municipality	41	Pierīga
2	Carnikava Municipality	41	Pierīga
3	Mārupe Municipality	38	Pierīga
4	Saulkrasti Municipality	38	Pierīga
5	Garkalne Municipality	35	Pierīga
6	Ķekava Municipality	35	Pierīga
7	City of Rīga	35	Rīga
8	Sigulda Municipality	35	Pierīga
9	Alsunga Municipality	34	Latgale
10	Ādaži Municipality	34	Pierīga
11	Babīte Municipality	34	Pierīga

Source Summary made by the authors using data of the CEC of Latvia

(EU). Participation in elections to the EP is granted to those people who have the right to vote and who are the citizens of Latvia or of other EU Member States with residence in Latvia, provided that information about them is included in the Population Register of Latvia. Also, Latvian citizens who live abroad are entitled to participate in the elections to the EP. The right to vote in Latvia is granted to those who have reached the age of 18. To record the people who have the right to vote in the EP elections, the electoral register is applied, and voting is carried out using polls. Each voter was registered for the elections in a particular electoral district according to their registered place of residence. A total of 30 % of voters participated in the elections to the EP in 2014. In 38 of 119 Latvian local governments, the voter turnout exceeded 30 %: in two of these, it was 41 %, while the majority (81) of the local governments showed the voter turnout under 30 %, and in three of them—20 %. The local governments with the highest voter turnout, as well as their location according to the breakdown by statistical regions, are indicated in Table 11.3.

The highest turnout in the EP elections in 2014 was observed in Ikšķile (41 %) and Carnikava (41 %) municipalities (both are part of Pierīga region, Table 11.3). This is due to the combination of two factors. First, there has traditionally been higher voter turnout in both regions than the average data of the country at large. When calculating the average turnout of those who have the right to vote in the local and EP elections in 2009, as well as in the local elections of 2013 and EP elections in 2014, Ikšķile and Carnikava municipalities are the ones with the most active voters. An additional factor of turnout in the EP election possibly was the fact that there were four persons among the EP candidates with a registered place of residence in these regions, thus mobilising supporters of the candidates to be active.

11.3.4 Comparison of Voter Turnout in Local Elections of 2013 and in European Parliament Elections of 2014

Comparison of voter turnout in the local elections of 2013 and EP elections in 2014 in Latvia allows establishing a positive relevance (Fig. 11.5). Although correlation is weak (0.3431), still it is statistically significant as for the chosen 5 % confidence level (the critical r value with the observation number 119 is 0.1801).

Even though the turnout in the EP elections in Latvia was lower (30 %) than in local elections (46 %), still about 10 % points higher was the number of voters who claimed that they had participated in the EP elections *because of civic duty* (Marketing and public opinion research centre SKDS 2014). Thus, the analysis of elections to the EP could enable clearer identification of motives that facilitated civic awareness, not just formal affiliation to a particular administrative territory. This served as a basis for creation of two models where both of them contain the same explanatory variables, while the explanatory variable, voter turnout in local elections in 2013 and in the elections to the EP in 2014, differs.

Figure 11.5 containing a comparison of voter turnout in both elections using boxplots shows that voter turnout in elections to the EP in 2014 was considerably lower than that in local elections in 2013. This is demonstrated by both a difference of the average value of voter turnout by more than 15 % points and a fact that the 2nd and 3rd data quartiles do not overlap (Fig. 11.4).

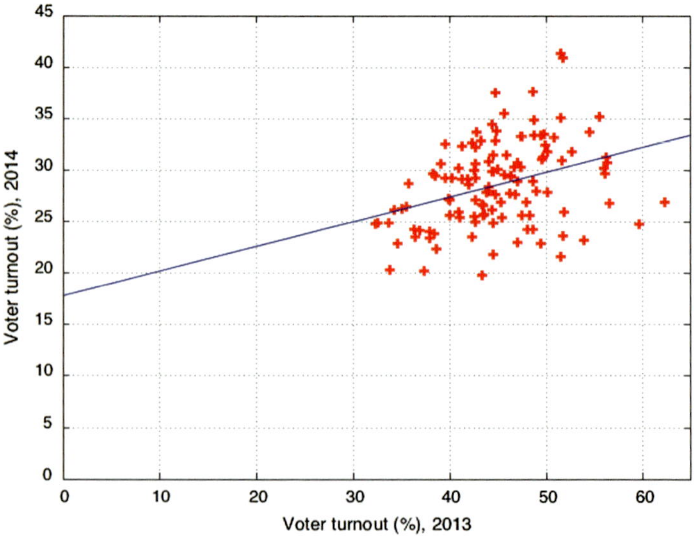

Fig. 11.5 Correlation between voter turnout in the local elections in 2013 and elections to the European Parliament in 2014. *Source* Calculations made by the authors, data of the CEC of Latvia

11.4 Socio-Economic Factors and Election Turnout in Latvia: A Model

11.4.1 Description of Variables Used in the Model and Justification of the Territorial Data

For the purpose of analysis, an econometric model was developed. First of all, a database was created, which included data of 22 variables from different official databases that are used when planning and organising the regional development policy of local government by the Latvian government, as well as the results of other participation studies carried out in 2012 and 2013. In the next stage, correlation among the variables was analysed in order to select the variables that explain voter turnout avoiding multicollinearity issues.

In the selection of the variables, the following conditions were taken into account:

- inclusion of as many *variables* as possible in the model that might explain voter turnout;
- avoiding inclusion of *explanatory variables* in the model that mutually closely correlate and may cause multicollinearity issues making credibility of the parameter assessment questionable.

After testing the mutual relations among the variables, it was found that there was a close connection among some of them. For instance, there is a close positive correlation exceeding 0.9 among *revenue from personal income tax per capita, average wage subjected to insurance contributions,* and *average age pension.* In order to reduce a possibility of multicollinearity, only the variable that correlates least with other variables from the database of those having high correlation has been kept.

After reviewing several model specifications and examining the assessments regarding multicollinearity (VIF test), the following six variables have been selected as explanatory variables:

- revenue from personal income tax in the budget of a local government per capita;
- the number of population (*its logarithmic value*);
- the number of inhabitants having higher education per 1000 inhabitants;
- the number of old-age pension beneficiaries per 1000 inhabitants;
- the proportion of the Latvian residents, percentage;
- the number of live births per 1000 inhabitants.

In the research, each of the six variables has been characterised, and cases that differ have been reasoned (extreme points).

11.4.2 Revenues from Personal Income Tax in the Budget of Local Governments Per Capita

Revenues from personal income tax in the budget of local governments per capita characterise to a great extent both the size of regular income of residents and the arising welfare differences in various administrative territories of Latvia. The personal income tax (PIT) applies to the income earned by a physical person. According to the Latvian legislation, division of revenue from the personal income tax is as follows: 80 % is constituted by a part deducted in budgets of local governments, while 20 % in the national budget. Hence, the personal income tax is a significant source of revenues for local governments, constituting almost 85 % of all tax revenues in the budget of local governments, and is closely related to performance of the local governments. All factors, be it individual, economic, or political, that reduce the revenue from PIT affect, to a great extent, the budgets of local governments, the execution of their functions, and territorial development. The personal income tax is distributed according to the registered place of residence of the income recipient. Revenue of PIT in local governments in Latvia, in 2014, was planned to be EUR 495 per capita, on average per year. Comparison of income data as for the personal income tax per year among local governments allows establishing that the lowest income from this tax is earned by local governments in Latgale region, while the highest by local governments of Pierīga region. The lowest revenue from the personal income tax, EUR 192, was in Riebiņi Municipality (Latgale region), while the highest, EUR 907, in Garkalne Municipality (Pierīga region).

Personal income tax revenues per capita are much higher in Pierīga region. Pierīga region is characterised by high mobility—its residents mainly arrive to work to other local governments—mainly to Rīga. The income tax is distributed according to the residence but not according to the work place of the income recipient, thus ensuring Pierīga region, which was used to be called a sleeping district of Rīga, a comparatively high revenue of PIT in local government's budgets (Fig. 11.6).

The system of distributing the PIT has arisen many discussions on unequal distribution of the revenue among municipalities ensuring a higher budget income for the region of Pierīga than to the local governments that provide work places to their residents.

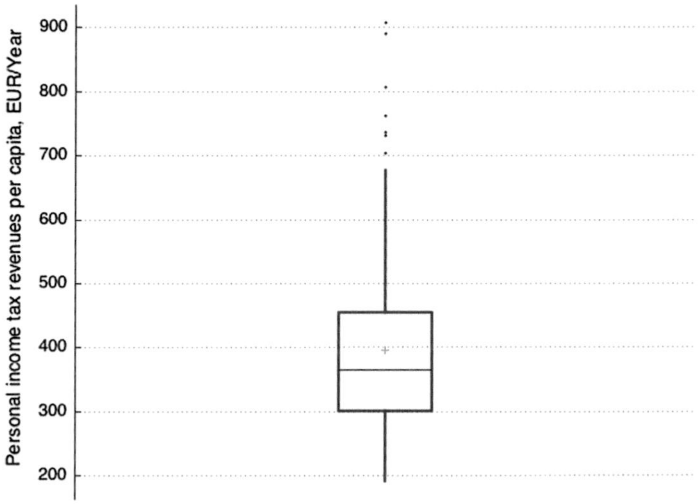

Fig. 11.6 Distribution of municipalities by personal income tax revenues per capita in Latvia in 2013. *Source* The boxplot made by the authors using the calculation on revenue from the personal income tax in the budgets of local governments per capita available in the regulations of the Cabinet of Ministers of the Republic of Latvia (2013)

11.4.3 Number of Population (Its Logarithmic Value)

The number of inhabitants is the total sum of resident population, for whom the relevant administrative territory is the permanent place of residence. The structure of resident population is usually analysed by age groups—under the working age, at the working age, and above the working age—determined according to the giving age and retirement age provided in the legislation of the respective years. According to the data of the OCMA, used by the Latvian government to implement the financial policy, there are 2 million and 201,196 people in Latvia; according to the data of the Central Statistical Bureau (CSB), this figure is 177,371 lower.[2]

The boxplot (Fig. 11.7) confirms the fact that in Latvia, in regard to the number of population, there is an essentially different municipality—the capital Rīga. According to the data of the OCMA, there are 696,618 residents in Rīga, or 32 % of the whole Latvian population. Slightly over 4 % of the Latvian population live in Daugavpils, which is the second largest city of Latvia considering the number of population.

The actual number of population in local governments, in international participation studies, is often being used as a variable in order to check the correlations between active participation of population and the size of local governments.

[2]In both cases, the chapter uses data referred to 1 January 2013.

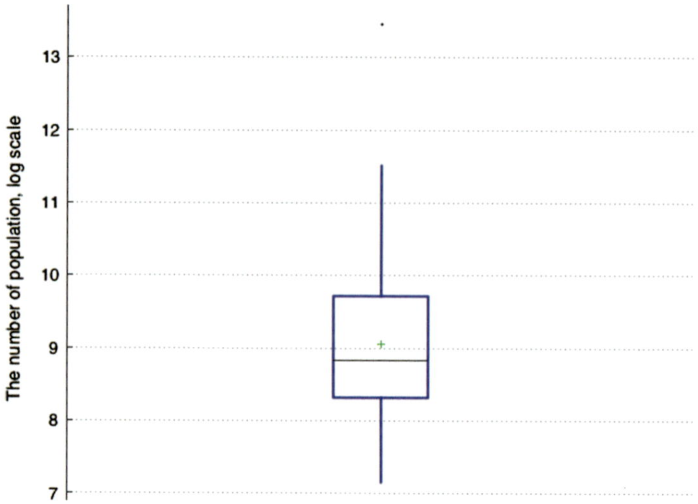

Fig. 11.7 Distribution of municipalities by the number of population (log scale) in Latvia in 2013. *Source* A boxplot developed by the authors using the OCMA data about the number of population as of 1 January 2013

11.4.4 The Number of Population with Higher Education Per 1000 Inhabitants

According to the survey data on labour force, made by the CSB, in 2012, 13 % of the inhabitants with higher education, 36 % with vocational or vocational secondary education, 33 % with general secondary education, and 9 % with elementary education were searching for a job. Acquired higher education has reduced the possibility to become unemployed almost three times, compared to job seekers who have vocational or vocational secondary education (CSB Website Statistics Database). Data available to the CSB characterise also the correlation between gross salary and levels of education. Employees with higher education receive higher gross salary per year (EUR 907) than those having obtained a lower level of education—vocational secondary (EUR 538) or general secondary education (EUR 518). The salary of the employees with higher education is higher than the average monthly gross salary in the state per year (EUR 665). This means that people with higher education are the value added to any local government from the point of view of both human capital resource and economic capacity. On average, 20 % of Latvian population have higher education (CSB Website Statistics Database). The biggest share is in Rīga (28 %) and municipalities of Pierīga region: Garkalne (32 %), Mārupe (29 %), Carnikava (29 %), Ikšķile (28 %), Babīte (27 %), and Ādaži and Ķekava (25 %, CSB Website Statistics Database). Municipalities of Pierīga region, when compared to those elsewhere, stand out in regard to the number of population with higher education (Fig. 11.8).

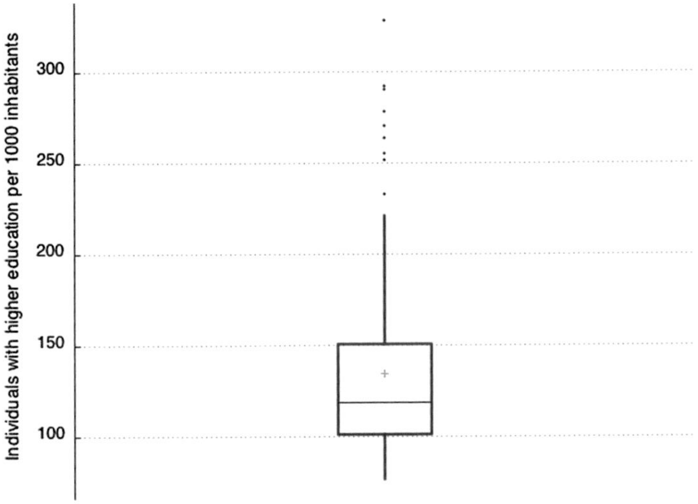

Fig. 11.8 Distribution of municipalities in Latvia by individuals with higher education per 1000 inhabitants in Latvia. *Source* A boxplot developed by the authors using the CSB's data on higher education

11.4.5 Old-Age Pension Beneficiaries Per 1000 Inhabitants

The number of old-age pension beneficiaries characterises the number of the persons who have reached the statutory age of retirement, provided that their insurance period has been at least 10 years. Till 31 December 2013, the retirement age in Latvia for both women and men was starting from 62 years, and with each coming year, it is increased by 3 months until it reaches the level of 65 years. The age pension is received by approximately 22 % of the Latvian population, according to data of the State Social Insurance Agency (SSIA). The smallest relative number of old-age pension beneficiaries is in municipalities of Pierīga region, whereas the highest one is in municipalities of Strenči, Ērgļi, Mazsalaca, and Rucava.

In the municipalities of Ādaži, Babīte, Garkalne, Ķekava, and Mārupe in Pierīga region, the relative number of old-age pension beneficiaries is smaller than in other municipalities (Fig. 11.9).

It cannot be unambiguously stated that there is a correlation among the number of old-age pension beneficiaries and the number of population at and above the giving age in all local governments. However, in the case of Pierīga, in the municipalities with the lowest number of old-age pension beneficiaries—in municipalities of Mārupe (12 %), Ādaži (14 %), Garkalne (14 %), Babīte (15 %), Stopiņi (16 %), and Ķekava (17 %)—the number of population at giving age is higher, namely in Ādaži and Garkalne (above 65 %), Babīte, Mārupe, and Stopiņi (almost 65 %), and Ķekava (64 %). The analysed statistical data explain why the numbers of old-age pension beneficiaries in the municipalities of Ādaži, Babīte,

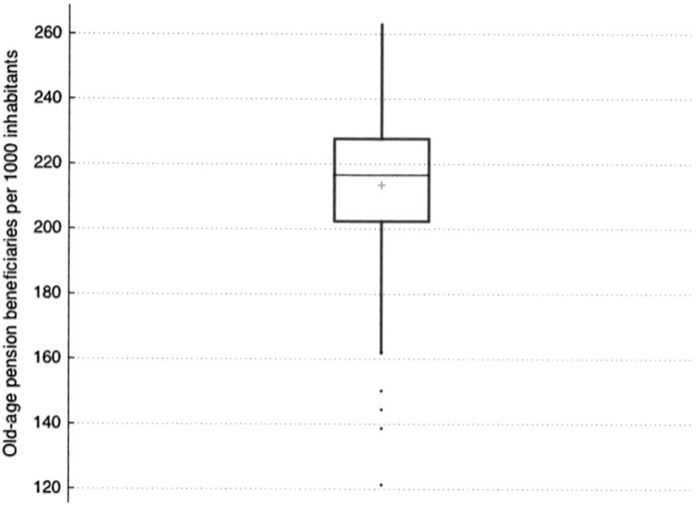

Fig. 11.9 Distribution of municipalities by old-age pension beneficiaries per 1000 inhabitants in Latvia in 2013. *Source* A boxplot developed by the authors using the SSIA data on the number of old-age pension beneficiaries

Garkalne, Ķekava, Mārupe, and Stopiņi are smaller than in other local governments; therefore, data on these municipalities are not considered to be untypical (outliers).

11.4.6 Proportion of Latvians

The proportion of Latvians as a variable is being used to explain the differences in *participation* of population between the inhabitants of the basic ethnicity and of other ethnicity. According to the CSB data, at the beginning of 2013, there were 61 % Latvians and 39 % people of other nationalities living in Latvia, and among the latter, the biggest ethnic group includes Russians—26 %, Belarusians—4 %, Ukrainians, Poles—2 %, Lithuanians, Jews, and representatives of other ethnicity (CSB Website Statistics Database). As to the breakdown by statistical regions, the greatest numbers of Latvians reside in Rīga region (24 %), Pierīga (21 %), Kurzeme (16 %), followed by Vidzeme, Zemgale (14 % in each), and Latgale (11 %). When relating the number of Latvians against the total number of population in a particular statistical region, the highest proportion of Latvians is in Vidzeme—86 %, Kurzeme—76 %, Pierīga—71 %, and Zemgale—70 %, whereas in two regions—Rīga and Latgale (45 % in each of them)—there are less Latvians than people of other ethnicity (CSB Website Statistics Database).

From the point of view of ethnic composition, the smallest proportion of Latvians (Fig. 11.10) is found in Daugavpils (according to CSB, 19 %), Daugavpils

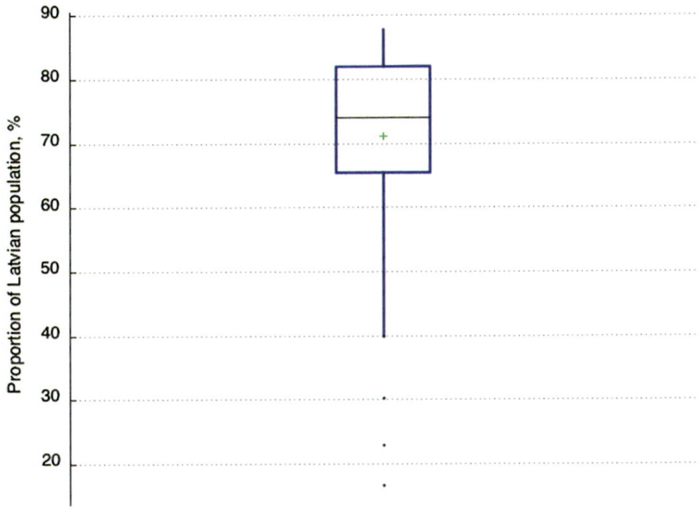

Fig. 11.10 Distribution of municipalities by proportion of Latvian population in Latvia in 2013. *Source* A boxplot developed by the authors using the CSB data about the number of population at the beginning of 2013

Municipality (according to CSB data, 34 %), and Zilupe Municipality (according to CSB data, 25 %). The above-mentioned municipalities have a higher proportion of other ethnic groups.

The biggest ethnic group living in the above-mentioned municipalities are Russians. The proportion of this nationality makes more than a half of the population in Daugavpils (50 %), Daugavpils Municipality (54 %), and Zilupe (55 %).[3] There is a large proportion of Poles and Belarusians in both Daugavpils city and Daugavpils Municipality. In Zilupe, there is a less distinct presence of other ethnic groups—including 3 % of Poles and 2 % of Ukrainians residing there.

The proportion of Latvians is <50 % in Jūrmala, Rēzekne, and municipalities of Krāslava, Olaine, Salaspils, and Viļāni, whereas municipalities with the highest share of Latvians—nearly 90 % and more—include Alsunga, Dundaga, Lubāna, Pārgauja, Roja, and Smiltene.

11.4.7 The Number of Live Births Per 1000 Inhabitants

The number of live births per year with one of the parents being a permanent resident of the Republic of Latvia in current demographic situation is used as an important territory development indicator. As from 2000, the number of births has

[3]Data of the CEC of the RL at the beginning of 2013.

been recalculated including the children who were born abroad when a mother (a permanent resident of Latvia) was abroad for a short period of time.

Birth is very closely related to objective characteristics of the national economic situation, as well as to other indicators of population behaviour and practical action. In the elaboration of the research, the authors used data on the number of live births in 2012. The average number of births in local governments of Latvia in this year was almost 9 births per 1000 inhabitants. In Mārupe Municipality, the number of births nearly twice exceeded the average number of births in Latvia—17 births per 1000 inhabitants. Comparatively higher birth rates per 1000 inhabitants are in the municipalities of Pierīga region—Stopiņi, Ādaži, Ikšķile, Ķekava, Garkalne, and Sigulda. Whereas the smallest number of children, 5 per 1000 inhabitants were born in Aglona Municipality (Latgale region). There are low birth rates also in other municipalities in Latgale (Kārsava and Riebiņi municipalities), Vidzeme (Varakļāni Municipality), and Kurzeme (Nīca and Mērsrags municipalities).

The highest birth rate is in Mārupe Municipality (Pierīga region, Fig. 11.11). This can be explained by several factors. The number of inhabitants in Mārupe Municipality has been growing year by year (State Regional Development Agency 2013).

Birth rate is affected by the age structure of population. 23 % of Mārupe population are under the giving age (younger than 15 years), 65 % are of the giving age, and 12 % above the giving age (CSB Website Statistics Database). According to the data of the CSB on 1 January 2013, the number of population under the giving age per 1000 inhabitants was 351, which is significantly higher than the overall numbers in Latvia, 229 (CSB Website Statistics Database). This means that

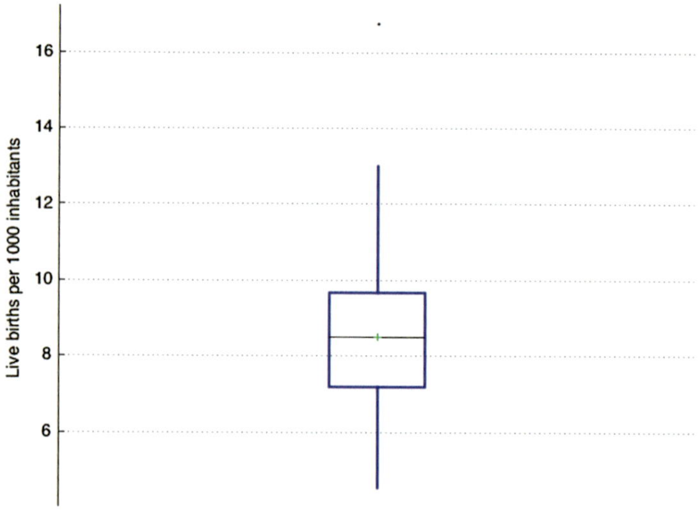

Fig. 11.11 Distribution of municipalities by live births per 1000 inhabitants in Latvia in 2013. *Source* A boxplot developed by the authors using the CSB data on the number of births

the administrative territory of Mārupe Municipality is chosen as a place of residence by relatively young people or families with children.

Although population's opinions differ concerning the role of a local government in increasing of birth rate, still the purposeful policy of Mārupe Council, which was implemented to attract new families by developing corresponding construction works and ensuring the necessary infrastructure to families with children in the local government, has justified itself. In Mārupe Municipality, the proportion of women aged 20–34 is 26 %, which is the highest proportion of young women in the age structure of population in Latvian municipalities, including in Pierīga region (20 %, CSB Website Statistics Database). In the case of Mārupe, the proximity of capital city Riga is a significant advantage, which allows ensuring workplaces to its residents.

Third, the birth rate is affected also by the state of population welfare. It is characterised by both the revenue from personal income tax per capita in the budget of a local government and the index of territorial development level showing such level in the relevant year. As for Mārupe Municipality, compared to the whole group of Latvian municipalities, the indicators in all the above categories are among the highest. Revenue from personal income tax per capita in 2014 for Mārupe Municipality was planned as the second highest revenue (EUR 890) following Garkalne (EUR 907, Pierīga region), and there was the highest index of territorial development level both in 2012 and 2013 (Cabinet of Ministers of the Republic of Latvia, Regulations 2013). Since for the untypical data of Mārupe Municipality socio-economic interpretation can be given, they are not discarded.

11.4.8 Description of the Theoretical Model

To assess the impact of various factors on voter turnout in elections of 2013 and 2014, the logistic regression was used. In the models, the following variables were tested (expected signs of effect on voter turnout are given in parenthesis):

- revenue from personal income tax per capita (+);
- population size (−)
- population with higher education per 1000 inhabitants (+);
- old-age pension beneficiaries per 1000 inhabitants (+);
- live births per 1000 inhabitants (+).

In local governments with a higher proportion of Latvians, a voter is more reachable. Such an assumption is made by the authors based on the results of the survey of Latvian population in 2012 where in a number of participation-related aspects (trust in public institutions, active action to involve in protection of ones' interests in a local government, and so on), Latvians provided a higher assessment than people of other nationalities (Seimuskane and SKDS 2012). This is the ground for the assumption that with the growing proportion of Latvians, the voter participation could grow, too, i.e. there could be a "+" sign for the beta coefficient.

When analysing institutional and non-institutional forms of participation, it was established in a range of international studies that younger people prefer non-institutional participation forms, while the elderly ones prefer institutional forms (Putnam et al. 1993; Zukin et al. 2006; Henn et al. 2002). The authors studied the possibility of such correlation also in Latvia. Although the study results failed to demonstrate clearly dominant usage of institutional or non-institutional forms, election as an institutional participation form is preferred by older people (Seimuskane et al. 2015). This substantiates the assumption that increase in the number of old-age pension beneficiaries would also lead to an increase in voter participation.

11.4.9 Evaluation of the Model

Estimated parameters of logistic models and related statistics are shown in Table 11.4.

Overall, the models are statistically significant, since in both cases the calculated F value is large enough for standard significance levels of 1, 5, and 10 %, so that the authors can reject the hypothesis that all coefficients are not simultaneously significantly different from zero.

In order to interpret the obtained parameter estimates given that for the logistical function marginal effects are not constant:

Table 11.4 Estimates of logistic regressions of voter turnout in local elections in 2013 and EP elections in 2014

Dependent variable	Voter turnout in the local elections of 2013	Voter turnout in the EP elections in 2014
	β_i (p value)	
Constant (β_0)	−0.2126 (0.5831)	−2.4716** (0.0000)
Proportion of Latvians (% of all inhabitants)	−0.0701 (0.6559)	0.4737** (0.0000)
Personal income tax per capita	−0.0002 (0.6217)	0.0007** (0.0003)
The number of live births per 1000 inhabitants	0.0070 (0.6003)	0.0172* (0.0161)
The number of inhabitants having higher education per 1000 inhabitants	0.0033** (0.0012)	0.0019** (0.0005)
The number of old-age pension beneficiaries per 1000 inhabitants	0.0046** (0.0001)	0.0039** (0.0000)
Number of inhabitants (logarithmic)	−0.1487** (0.0000)	−0.0347** (0.0055)
R^2	0.3838	0.7543
$F(6.112)$ (p value)	11.6246 (0.0000)	57.3129 (0.0000)
Normality of residual (errors are normally distributed) χ^2 (2) (p value)	3.0150 (0.2215)	2.8598 (0.2393)

*Significance at the 5 % level; **significance at the 1 % level

Source Developed by the authors

(1) independent variables of estimated regression equations are replaced by their mean values and the "average" Y are calculated:

$$\hat{Y}_{2013} = \frac{1}{1 + e^{-(-0.2126 - 0.0002x_1 - 0.1487x_2 + 0.0033x_3 + 0.0046x_4 - 0.0701x_5 + 0.0070x_6)}} \text{ and } \quad (11.1)$$

$$\hat{Y}_{2014} = \frac{1}{1 + e^{-(-2.4716 + 0.0007x_1 - 0.0347x_2 + 0.0019x_3 + 0.0039x_4 + 0.4737x_5 + 0.0172x_6)}}, \quad (11.2)$$

where designation of independent variables x_i ($i = \overline{1;6}$) is given in Table 11.4;

(2) "average" values of independent variables are increased or reduced by the amount so that the dependent variable will change by one percentage point.

Following the above algorithm, the authors obtained the following results.

Signs for parameters of statistically significant coefficients correspond to those that were theoretically expected. In the model of elections in 2014, all coefficients are statistically significant, with the expected theoretical signs, whereas in the model of elections in 2013, the coefficients of the variables—population with higher education per 1000 inhabitants, old-age pension beneficiaries per 1000 inhabitants, and the population size—were statistically significant and having the correct theoretical signs.

Chi-square values for both models are low enough not to reject the hypothesis about non-normal distribution of residuals with any of standard significance levels (10, 5, or 1 %).

As a result of the study, the authors concluded that the number of people with higher education is a significant factor both in local elections and in elections to the EP. This means that individual factors of an individual's status and behaviour determined by knowledge, skills, and civic awareness are of greater importance in participation motivation of an individual. The level of voter turnout in elections to the EP is affected by the relative number of live births, suggesting that the participation level is higher in the local governments where there are more favourable preconditions to child development. The number of births in families closely correlates with income and higher education of inhabitants. An ensured level of welfare and education gives an opportunity to evaluate the importance of elections to the EP. Elderly, who are the old-age pension beneficiaries, are more active users of institutional forms. Life experience, civic discipline, and active life position motivate participation in local and EP elections.

The obtained results allow concluding that the size of the number of inhabitants is of higher importance in local elections rather than in elections to the EP. Changes in the number of inhabitants in local governments may affect voter turnout in local elections and to a smaller extent—in elections to the EP. This means that affiliation of a citizen to a specific administrative territory has its role in local elections but not in elections to the EP.

11.5 Conclusions

The following explanatory factors of the voter turnout in Latvia concerning features of a socio-economic status have been identified:

- The number of residents in a local government unit in Latvia is of a greater importance in local elections than in the elections to the European Parliament. With the increase in the number of residents in the local government, the voter turnout in local elections could tend to decrease, the less evident this connection would be concerning the elections of the EP. This means that affiliation of a citizen to a particular administrative territory is of importance in local elections rather than in elections to the EP;
- Higher education is a significant factor in promotion of the voter turnout, besides it is more significant in local elections in comparison with the elections to the EP. This means that individual factors of an individual's status and behaviour determined by knowledge, skills, and civic awareness are of great importance in participation motivation of the individual;
- The number of live births closely positively correlates with the income level and higher education level of the population. An ensured level of welfare and higher education gives a preference to assess the importance of elections to the EP.
- Older people, old-age pension beneficiaries, are more active users of institutional participation forms. Life experience, civic discipline, and an active life position motivate them to participate in local elections and elections to the EP.

The data obtained on voter turnout and on the social structure of population in local governments in Latvia allow concluding that the socio-economic factors have even importance in both type of elections—local and EP. In addition, voter turnout in the local elections is also affected by local contextual factors.

However, there is a very close link between certain socio-economic factors of the population of a local government unit and the voter turnout in elections to the European Parliament in Latvia. Almost all local governments with the highest voter turnout in the EP elections are located in the statistical Pierīga region—these are the local governments with highest revenue from personal income tax in their budget, where there is the greatest proportion of population having higher education, the highest employment proportion, and the highest birth rate.

It is not simple to compare the influence of economic factors on voter turnout in Latvia with similar studies in other European countries, since there are distinctions among practising democratic traditions and also political participation procedures. Most feasibly, citizens of countries with a longer history of democracy will also happen to be more politically mature. However, it is possible to analyse the research results and seek for interconnections.

Viewing studies on voter turnout in various periods (The Citizens' Audit in Britain (2004), European Social Survey (2006/2007)) and comparing them with the results of Latvian research, a similar tendency can be distinguished—education is a significant factor in facilitating voter turnout in both the local elections and European Parliament elections in Latvia.

The authors' research results confirm also a certain positive connection among the welfare level of the voters, birth rate data, occupation status, and activity of their political participation, which are significant factors in European studies. However, in Latvia, this connection is more pronounced in European Parliament elections, while in local elections there dominate contextual, although sometimes hardly explainable factors characteristic of the particular administrative territory, which is an important finding for studies on urban governance in Europe and in other regions of the world.

References

Almond GA, Verba S (1989) The civic culture: political attitudes and democracy in five nations. Sage Publications, London

Bendor J, Diermeier D, Ting M (2003) A behavioral model of turnout. Am Polit Sci Rev 97 (2):261–280

Brady H (1999) Political participation. In: Measures of political attitudes. Academic Press, San Diego, pp 737–801

Cabinet of Ministers of the Republic of Latvia (2013) Regulations on revenues of local government finance equalization funds and their distribution order Nr:1433. Accessed 4 April 2015. http://likumi.lv/doc.php?id=263003

Denters B, Goldsmith M, Ladner A, Mouritzen PE, Rose LE (2014) Size and local democracy. Edward Elgar Publishing, Cheltenham

Duch R (1998) Participation in the new democracies of Central and Eastern Europe: cultural versus rational choice explanations. In: Barnes SH, Simon J (eds) The postcommunist citizen. Erasmus Foundation and Institute dor Political Science of the Hungarian Academy of Sciences, Budapest, pp 195–228

European Commission (2012) Standard Eurobarometer 78. Accessed 4 April 2015. http://ec.europa.eu/public_opinion/archives/eb/eb78/eb78_en.htm

European Commission (2013) Standard Eurobarometer 80. Accessed 4 April 2015. http://ec.europa.eu/public_opinion/archives/eb/eb80/eb80_anx_en.pdf

European Commission (2014) Standard Eurobarometer 82. Accessed 4 April 2015. http://ec.europa.eu/public_opinion/archives/eb/eb82/eb82_anx_en.pdf

Feddersen TJ (2004) Rational choice theory and the paradox of not voting. J Econ Perspect 18 (1):99–112

Geys B (2006) Explaining voter turnout: a review of aggregate-level research. Electoral Stud 25 (4):637–663

Henn M, Weinstein M, Wring D (2002) A generation apart? Youth and political participation in Britain. Br J Politics Int Relat 4(2):167–192

Hooghe M, Quintelier E (2013) Political participation in Europe. In: Keil IS, Gabriel O (eds) Society and democracy in Europe. Routledge, London, pp 220–243

Keil SI, Gabriel OW (eds) (2013) Society and democracy in Europe. Routledge, London and New York

Marien S, Hooghe M, Quintelier E (2010) Inequalities in non-institutionalised forms of political participation: a multi-level analysis of 25 countries. Polit Stud 58(1):187–213

Marketing and public opinion research centre SKDS (2014) Survey of Latvian residents—study of voter attitude 2014. CEC of Latvia. Accessed 15 June 2015 http://cvk.lv/pub/upload_file/Petijumi/atskaite_CVK_062014.pdf

Pacek AC, Pop-Eleches G, Tucker JA (2009) Disenchanted or discerning: voter turnout in postcommunist countries. J Polit 71(2):473–491
Parliament of the Republic of Latvia (1994) The City Council and Municipality Council Election Law
Parliament of the Republic of Latvia (2004) Law on Elections to the European Parliament
Parliament of the Republic of Latvia (2010) Law on Separation of Roja Municipality and Launch of Operation of the Newly Established Municipalities
Pateman C (1970) Participation and democratic theory. Cambridge University Press, Cambridge, p 119
Putnam RD, Leonardi R, Nanett RY (1993) Making democracy work: civic traditions in modern Italy. Princeton University Press, Princeton
Rosenstone SJ, Hansen JM (2003) Mobilization, participation and democracy in America. Longman, New York
Sartori G (1970) Concept misformation in comparative politics. Am Polit Sci Rev 64(4):1033–1053
Schlozman KL (2002) Citizens' participation in America: what do we know? Why do we care? In: Katznelson I, Milner HV (eds) Political science: state of the discipline. W.W.Norton & Company, New York, pp 431–461
Seimuskane L, Marketing and public opinion research centre SKDS (2012) Survey of Latvian residents—assessment of citizens' participation and democracy in local governments in Latvia
Seimuskane L, Vilka I, Brekis E (2015) Institucionālo un neinstitucionālo līdzdalības formu salīdzinājums pašvaldībās Latvijā. Latvijas intereses Eiropas Savienībā 2:54–64
Smith M (2009) The inequality of participation: re-examining the role of social stratification and post-communism on political participation in Europe. Sociologický časopis/Czech Sociol Rev 45(3):487–517
Stoker G (2006) The decline of citizen engagement? In: Why politics matters making democracy work. Palgrave Macmillan, New York, pp 87–102
Van Deth JW (2001) Studying political participation: towards a theory of everything? Mobilisation, Organisation and Participation via new ICTs, p 19
Verba S, Schlozman KL, Brady HE (1995) Voice and equality: civic voluntarism in American politics. Harvard University Press, Cambridge
Vilka I (2013) Administratīvi teritoriālais iedalījums un pašvaldības. In: Stradiņš J (ed) Latvieši un Latvija. III sējums. Atjaunotā Latvijas valsts. Latvijas Zinātņu akadēmija (Latvian Academy of Science), Riga, pp 214–242
Vilka I, Brekis E (2013) Comparing local government elections in Latvia and other European countries. Soc Res 3(32):5–16
Vilka I, Seimuskane L (2012) The evaluation of local government reform in Latvia from current perspective. In: The 20th NISPAcee annual conference public administration: east and west: twenty years of development, Ohrid, Republic of Macedonia, 23–26 May 2012, CD
VRAA (State Regional Development Agancy) (2013) Attīstības centru ietekmes areālu noteikšana un analīze. Plānošanas reģionu, republikas pilsētu un novadu pašvaldību attīstības raksturojums, Riga, p 151. Accessed 1 Oct 2014. http://www.vraa.gov.lv/uploads/documents/petnieciba/petijumi/regionu_attist_final_2012.pdf
Zukin C, Keeter S, Andolina M, Jenkins K, Delli Carpini MX (2006) A new engagement? Political participation, civic life, and changing Americas citizen. Oxford University Press, New York

Data Bases

Central Election Commission (CEC) of the Republic of Latvia www.cvk.lv
Central Statistical Bureau (CSB) of the Republic of Latvia www.csb.gov.lv
Office of Citizenship and Migration Affairs (OSMA) of the Republic of Latvia www.pmlp.gov.lv
State Regional Development Agency (SRDA) www.vraa.gov.lv
State Social Insurance Agency (SSIA) of the Republic of Latvia www.vsaa.gov.lv

Chapter 12
Making Informed Citizens in Local Direct Democracy. What Part Does Their Government Perform?

Werner Pleschberger

Abstract This chapter has two main parts. First, it starts with a short presentation of the well-established theorem of the uninformed voter, and argues that local direct democracy inhibits systemic incentives for a voter to become more open for information, and for a local government to inform the voter. Due to the information asymmetry, local government is required to actively make the local voter better informed to decide competently on a ballot measure. Local government further has to adopt practical standards to fulfill the task, such as appropriate length, comprehensiveness, objectivity, and political neutrality of voter information. Second, this chapter describes three cases of official local voter information, how they are regulated by law and work in practice. The poor regulation and practice of voter information of the City of Vienna gives the party politics and public officials a very free hand to manipulate the task of informing the voter in their own interests. The Austrian City of Bregenz represents a moderate example, whereas the City of Los Angeles has the merit of a high developed approach of voter information. In all cases, the politicization of voter information is ever present. The gap between practical standards and the real information environment can preclude voters making informed decisions at the polls and indirectly through that also on urban governance.

Keywords Local government · Local direct democracy · Voter information · Urban governance · Austria · USA

12.1 Introduction

It is widely expected that a typical voter in a representative democracy should be a political sophisticate for his own sake and the sake of democracy. Accordingly, the voter should have sufficient knowledge of the institutions, stakeholders, and the

W. Pleschberger (✉)
University of Natural Resources and Applied Life Sciences, Vienna, Austria
e-mail: werner.pleschberger@boku.ac.at

© Springer International Publishing Switzerland 2017

C. Nunes Silva and J. Buček (eds.), *Local Government and Urban Governance in Europe*, The Urban Book Series, DOI 10.1007/978-3-319-43979-2_12

pressing issues of the polity, knowing the alternative solutions of the political elites and what their consequences may be (Berelson et al. 1954: 308). The postulated political knowledge is verifiable (Bendor and Bullock 2008: 3; Rapeli 2014: 8–12; Somin 2013: 9). There are some scholars for whom the voter is a political illiterate (Sartori 1992). The empirical research in the USA has long documented the low level of knowledge that the American voters hold on the simplest facts about the institutions or political officeholders of their national political system (Elmendorf and Schleicher 2013: 371–373; Somin 2013: 17–24). This finding has not really changed in the long run, despite huge financial investments in the education system (Bendor and Bullock 2008: 4). This is a finding that has also been verifiably found elsewhere in the world (Rapeli 2014: 30).

In the sphere of direct local democracy and in the role of a local lawmaker, the voter has to decide on a measure under their own agency. They should be able to make an informed decision, as they are acquainted with the facts, major dimensions and their interrelationship, and the relevant contexts of a measure on the ballot. Issue-specific knowledge is at the crux of political knowledge. Knowledge of the legal and fiscal context of a new measure is also crucial in order to achieve a notion of the implied policy change and the fiscal implications of a measure. The cognitive requirements of direct local democracy nourish skepticism about whether the voter can meet the expectations to make an informed decision (Elmendorf and Schleicher 2013: 390). Several findings show few local voters do not know the active political heads nor do they have profound knowledge of local issues (Garrett and McCubbins 2008: 51). The suggestion that the local lawmaker is not fully informed, is a generous understatement (Lupia and Matsusaka 2004). A valid objection to the expectation of knowing, it is that they would not be so minded to read policy-specific information about a measure (Warner 1995: 81; Eule 1990: 1516, 1991: 749). In contrast, controversial issues or personal emotional connections with an issue can stimulate the interests of the voter, to take a closer look at the content details or effects of a proposition (Bullock 2011: 497).

Anyway, in light of the politico-economic approach the informed voter is beyond a rational expectation (Fiorina 1990: 335). The rational voter normally avoids using personal energy, time, and other costs in the procurement, analysis, and evaluation of information. It is for them rational behavior to keep ignorant of information (Popkin 1994: 13–15). They cannot be sure whether investment in the procurement of information would be worth it for themselves. Even if they knew the individual consequences of a political choice for themselves, they even do not know the collective consequences of that choice (Buchanan 1954: 355). In a collective decision, an informed vote may carry as little or as much worth as an uninformed vote. Even if all remember to participate in an election, it is supposed only a few would have their voices represented (Downs 1957: 253). The larger the electorate is, the less an individual voice is represented, which is a systemic disincentive for voters to inform themselves. Contrarily, there are some residual incentives for voters to get informed and to participate in an election, like the fundamental valuation of democracy over that of dictatorship (Downs 1957: 274).

We argue more in general, that the institutional configuration of direct local democracy is a system of incentives, (a) for the voter to acquire appropriate political knowledge and (b) for the local administration, albeit in their view, to take action with informed consent.

More specifically, we argue that local government has the task of informing voters about the measure to be consented to. In light of the principal agent model it is the local administration who is the agent, and is acting on behalf of the voters, who are the principal. The administration agent can use his resources to collect and forward information to meet the goals and wishes of the electorate. The agent knows more than the typical voter about issues, facts, solutions, and so on. The agent can pursue own goals, which depart from the objectives of the electorate. Reasons provided in a voter guide may simply be lies and not useful information to voters, like marketers may offer reasons for the consumer to buy their stuff (Garrett and McCubbins 2008: 47). In order to avoid agency costs, the legal regulation of voter information is necessary. This should be as specified by the voters and respected by the administration.

Above all the voter information about a measure on a ballot is a practical exercise, whose design and communication will determine the success or failure of a ballot, something which researchers, with few exceptions do not appreciate (Cronin 1989; Baker 1991; Dubois and Feeney 1998; Garrett 2001; Garrett and McCubbins 2008; Burnett et al. 2010). It would be illusory and ineffective a trusty local administration to expect to inform all voters comprehensively about a measure (Lupia 1994). According to the concept of pragmatic democracy, the realistic goal of information is the normal Gaussian distribution of knowledge among the voters (Delli Carpini and Keeter 1996: 170–178; Rapeli 2014). Many concrete questions arise when getting closer to the implementation. In which phase of the initiative process should the voters be informed officially? Which public official is to take which task, and accordingly is responsible for the information? What practical standards shall guide the voter education? How to design the information booklets sent to the voters directly? How is the ballot question to be worded and, what information will be presented on the ballot paper directly? (Garrett 2001: 160).

The remaining parts of this chapter are organized as follows. In Sect. 12.2, we describe the information environment surrounding local direct decisions of propositions based on different streams of literature in theoretical and practical perspectives. Section 12.3 is devoted to the analysis of three practical cases of voter information which will help us to get an understanding how it works in the local political reality and therefore how it may affect local government and urban governance. This chapter finishes with concluding remarks in Sect. 12.4.

12.2 The Information Environment Surrounding the Voter in Direct Democracy

12.2.1 Systemic Information Incentives for Voters

A direct democratic process is commonly binary in nature. The voters want the measure or they do not want it. They make a simple choice on very often complex policy issues with long-term consequences, namely between the promise of change and the status quo. Another possibility is the decision between several explicitly made policy alternatives and their trade-offs (Elmendorf and Schleicher 2013: 390; Tolbert and Smith 2006: 26). When the voters are to decide on a common measure, the period between initiation and decision is shorter than a general election period. The smaller a local electorate is the stronger the impression is to an individual voter that their voice can count. Small electorates make it worthwhile for a citizen to be informed to reap prestige or to avoid social isolation (Benz and Stutzer 2004: 34). The voters can expect that a substantial measure on ballot may change the local life (Reilly and Walker 2010: 226; Pommerehne 1978: 258; Pommerehne and Schneider 1978: 395), because they do not decide over expectations and promises of competing parties or candidates (Somin 2013: 105).

Since any substantial measure on which the voter decides will have some fiscal implications, there is a specific self-stimulus for the voter in the role of a taxpayer to acquire knowledge about the likely revenue burden imposed on (Pommerehne and Schneider 1978: 396; Benz and Stutzer 2004: 56). As Tolbert et al. (2003: 32–34) suggest, citizens who more often are exposed to ballot initiatives, statistically speaking at minimum in the year of study are better informed. However, the impact of direct democracy heavily depends on the broader electoral context and the practical details of the voter information.

12.2.2 Information Asymmetry Between Voters and Their Local Governments

Since substantive issues, at least ideally, should predominate in a direct democratic decision, the collection, selection, and analysis of information are required and a first decisive step (Downs 1957: 209). Such a decision imposes high information and opportunity costs on the voters, which they will not, and mostly cannot rise to reach accordingly even if they so desire (Kruse 2001: 144). The ordinary voters cannot bring to bear that good level of factual analysis on a measure. They are almost entirely dependent on external sources, to offering them helpful supporting material (Funk and Gathmann 2014: 14; Kirchgässner et al. 1999: 55).

Viewed realistically, a reasoning voter would have to move through a complex environment of informal and official sources to acquire general and issue-specific knowledge, whereas the last one “provide a flavor of the information that is likely available through other outlets” (Garrett and McCubbins 2008: 57). Some scholars suggest, citizens could get information from media outlets. If the media cover an issue, they occasionally report and load it with symbolic meanings (Schacter 1995: 131–138, 148). When searching for limited and simple information, voters can take other shortcuts or so-called information substitutes. If given a choice between candidates, the voter can resort to classical voting cues to make a choice (Center for Governmental Studies 2008: 13). Party labels, party affiliation, personal background, the appearance of a candidate, or their political success in the past is an informational substitute for making informed political choices (Somin 2013; Bullock 2011). In ballot issue campaigns such informational political proxies are not always available compared to general elections (Primo 2013: 115). It is suggested, that voters use shortcuts like personal endorsements in a limited way and the conditions are unclear under what conditions they do so (Burnett and McCubbins 2013).

Therefore, in the information environment, substantive “governmental generated sources” (Schacter 1995: 135) or “state-provided information” (Lupia 1994: 65) are of crucial importance for the voters to get better informed about a measure they are asked to decide (Hastings and Cann 2014: 119). Official information sources would be available plentiful. Local governments have access to a huge amount of data and factual information. The administrative infrastructure of agencies and other units produce technical information which mainly may meet their own needs and interests and are of little direct use to the general public (Downs 1957: 227). Only a little out of the total they make accessible to the voters (Downs 1957: 217). Local governments select official information from a much larger pot, and in compliance with interests, rules, principles, and taking into account the time or financial costs of information acquisition, which makes them partial and judgemental (Downs 1957: 212). Modern information technologies facilitate the cost-effective distribution of much and perhaps too much information to the wider public (Somin 2013: 287). How far official information contributes to the political sophistication of the voter may be, among other broad factors, a function of substantial quality, accessability, and frequency of exposure. The voters may appreciate the official statements about a measure to decide, because in their minds are not directly associated with a particular political position (Burnett et al. 2010: 318).

12.2.3 Political Interests of Local Governments

Local governments are neither perfectly competent nor politically disinterested. A local government and the political party in the background want to win the next election and seeks to gain advantage for their own cause (Burnett and McCubbins 2013: 1594; Theodore 2013). In every action, the local government is not totally disinterested, fully impartial nor perfect, even if they want to be (Kim 2010; Theodore 2013; Burnett et al. 2010).

Opposite to the self-interested voter, local governments are interested to articulate their positions in style and content, especially when a proposed measure is contradictory to their position (Benz and Stutzer 2004: 34). They have an opinion and want to bring about a result at every debate and to all measures, in instances most voters do not want (Garrett and McCubbins 2008: 57). Therefore, it manifests that local governments may use voter information as a tool to endorse their own endorsements, which may mean manipulation in many ways (Burnett et al. 2010: 322). Even competent and honest officeholders cannot deny their individual values and experiences when they draft an official information (California Commission on Campaign Financing 1992: 184). The palette of informing reaches from genuine information, to deliberate manipulation, misinformation, or indoctrination. The ballot wording can be a political emissary to kill an initiative proposition (Burnett and Kogan 2015). Persuasive information has little relation to the facts (Burnett and McCubbins 2013). Facts can be manipulated or to give prominence to a false impression of an issue (Downs 1957: 208). In the voter information, fiscal implications of a measure to decide on may be displayed correctly, exaggerated, understated, or remain as costs. Self-interested bureaucrats and politicians tend to exaggerate the benefits of a measure and systematically understate the cost to the taxpayers and voters while disregarding the actual voter preferences. The voters may wrongly assume that the costs fall later, onto the next generation or even the generation after next, and the welfare benefits can be consumed today. The fiscal illusion, the systematic misperceptions of the fiscal effects of a measure on the part of the ruling elites and the electorate, is feeding the understanding that a public good costs little to nothing or something will appear to cover the costs (Pommerehne and Schneider 1978).

12.2.4 Three Public Information Formats

Based on a very simplified procedural model of a direct democratic decision[1] on a measure (at least) three official and state-provided information formats are useful to inform the voters:

- A briefing document, describing in a concise and compact manner the title and summary of a requested measure, similar to a nutrition label (Burnett et al. 2010: 318). It would show the direct aspects of a requested measure. The significance

[1]The very narrow understanding of a direct democratic decision is a measure that is initiated and decided by the voters. For purposes of our study, we include when a local council has decided to ask the voters to decide in a binding or consultative referendum. Which form of direct democracy is applied, depends heavily on the local political context and on legal provisions in the city charters.

and the implied changes for the life of the citizens, for the provision of other services, as for the overall financial situation of local governments would be outlined. Citizens who are to be motivated to support a measure should know what is basically planned if they are to commit with responsibility. It is to be avoided, that they are motivated to sign up by overly brief or wrong slogans and incomprehensible speeches on the part of proponents (Cronin 1989: 208).
- A voter guide or pamphlet is a more comprehensive document, which describes the important aspects and contextual conditions and effects of a measure on the ballot. It would be to a wider and deeper level than the briefing document, and thereby would increase the complexity of the information. It could also contain non-neutral elements, as various political statements, in support of or against a measure. Logistical information would to be contained in the document (Garrett 2001: 171).
- A voting card is the third document. It should inform concisely and comprehensively about the title of the ballot and the ballot question, perhaps with a short preamble. This for some voters may be the first and last bit of information before their vote for or against a proposed measure (Gafke and Leuthold 1979; Garrett 2001: 167–168; Burnett et al. 2010: 318).

All documents should include explanations about the financial implications of a measure, in a detailed disclosure or in an aggregated indication.

- Procedural costs (Jakubowski et al. 1997: 83)
- Direct costs of the measure (planning costs, capital costs, costs for construction, operation and maintenance costs over an estimated service life, and so on). The longer the period calculated, the more uncertain would be the result of calculation.
- Opportunity costs, meaning a comparison at least between a first and second alternative. As with most discussions in the economy on costs, the indication of the opportunity cost is not always done with a figure. Here it is simply the next best alternative that is given in most decision-making situations. The voters should be able to choose their preferred alternative from a limited number of alternatives, ideally with the best cost–benefit ratio (Jakubowski et al. 1997: 82).
- Costs of receiving, processing, and using information on the part of the address or recipients of a measure
- Information and financial costs for the proponents of a measure.

The fiscal dimensions of voter information are, with few exceptions, given marginal consideration in the scientific literature (Jakubowski et al. 1997; Burnett et al. 2010: 320; Carter 2011: 315–321; Levinson and Stern 2010; Leib and Elmendorf 2012). Voters should be able to work out an accurate fiscal statement on a ballot. If they did want to rely on their own resources, they would have to invest, if this was at all possible, extremely high information and learning costs, more than to other facets of a measure (Garrett and McCubbins 2008: 40). The average voter is lacking greatly an understanding of the direct financial implications and consequences of a ballot box decision and is with a highest probability unable to get the budgetary perspective surrounding the measure (Theodore 2013). That harsh statement may in particular emphasize the lacking of contextual fiscal knowledge. What is the budgetary situation of a local entity that placed a measure on a ballot?. When municipal bonds are issued, is the exact mode of repayment specified and understandable for the voters (Levinson and Stern 2010: 690)? And are they informed of the financial situation of the bonding authority? The authority's bond rating can show, retrospectively, how the issuing organization have financially operated, which for the voters is easy to understand, useful, and trusted, because the information is given by a third party that has no direct interest in the bond measure (Garrett and McCubbins 2008).

Objective and understandable information about the financial implications of a measure, including the budgetary context, would clamp down on feeding the electorate the view that there is a free lunch somewhere in the world.

12.2.5 Four Practical Standards

When designing official voter information outlets, four practical standards should be born in mind by a local government:

12.2.6 Appropriate Length

The length of a document should encourage the voters, and not discourage them from paying attention to it (Garrett 2001: 172). A lengthy document can exceed the cognitive and temporal capacity of an average voter or even an interested and legally knowledgeable voter (Eule 1990, 1991). The type of the substantive issue, the number of actors involved that contribute to voter information can drive the length. The number of measures on a ballot at an election also affects the length.

12.2.7 Width Intelligibility

All formats of voter information should have understandable content for as many voters as can be (Reilly and Richey 2011: 59). Reilly (2013), Reilly and Richey (2011) stress the importance of the ballot complexity. The thematic field of a measure or fiscal information on the measure increases complexity and hinder popular understanding. Their study of 1200 ballot propositions at the state level showed that, at best, 25 % of voters could fully understand them. If several measures differ only in details, the voter can find it hard to recognize what content is to be voted on. Parameters such as sentence structure, number of syllables, and the length of sentences or use of foreign words affect intelligibility. There are natural limits on coping with technical jargon that is commonly used in expert circles. The goal to simplify the texts of documents for more intelligibility competes with the intention of legal and technical accuracy. Public institutions tend to weight the legal accuracy over simplicity.

12.2.8 Objectivity

According to the supposition of objectivity, all important substantive elements of a measure and its context conditions would be presented. Introducing a new measure, like a new municipal service, may lead to a tax increase, to cuts in another program, and so on. Plausible options and desirable as undesirable consequences should be indicated with an equal weight. It should be clearly understood that you cannot just write anything in the pure advocacy of a measure. As Levinson and Stern (2010: 729) write: "Instead of merely deciding whether or not to approve a program, issue bonds, or reduce taxes and fees, the voters would also have to make detailed decisions about cutting other programs and identifying funding sources to make up for reduced revenue."

12.2.9 Strict Neutrality

The supposition of neutrality implies that voter information contains no overt or covert bias in favor of a political or other interest. A biased wording of texts leads the electorate in the wrong direction (Kim 2010). Politically ambiguous terms are not allowed. Positive terms can make the electorate support a measure where, perhaps for reasons of neutrality, a descriptive term would be more appropriate.

There is no accepted method to measure the existence and extent of a political bias of information (Elmendorf and Spencer 2013: 521). But as reality shows terminology can quickly become a political issue. For business interests, it makes a difference whether in a document "major polluters" or "major sources of emissions" is written. A popular initiative for limiting municipal expenditure has an interest in that relevant documents which use simple terms such as "expenses" and "limit" rather than complicated explanations which technically reflect the facts in a legal or fiscal terminology.

12.3 Case Studies: Vienna, Bregenz, and Los Angeles

12.3.1 Methodological Remarks

It is the common wisdom of political science that institutions matter. According to March and Olsen (2008), political institutions correlate with a different agency performance of the system. As a set of rules, norm and operating procedures they at least partly impact the behavior of the political actors, and are not solely the result of utility maximization of involved actors. Institutions are a given reality, which cannot be changed arbitrarily. They are sometimes contested in a debate between insiders and outsiders, but if a change may occur this mostly reflect a limited adaptation to a new experience. Often institutions stay immune and do not react to the rate of change in the environment.

Thus, we broadly assert that local institutions matter at least partly and influence the performance of voter information in direct democratic decisions as provided by a local government. Comparing, some authors hypothesize that cities with directly elected mayors are more responsive to the views of their citizens than cities with a council–manager system (Tausanovitch and Warshaw 2014: p. 607). Vice versa, it can be assumed that an indirectly elected mayor is less visible, less partnership-oriented, and more open to a partisan mayoral policy.

The empirical part of this chapter investigates three practical examples of how city governments inform citizens about measures in the framework of our simple model of a direct democratic decision on a measure. The objects of analyses are three cities: Vienna, in Austria, which also holds Austrian federal state status; Bregenz which is located in the western Austrian federal state of Vorarlberg; and the City of Los Angeles in the US federal state of California. They represent different local institutional systems and perform very different approaches to practical voter information.

The basic institutional pattern of the local government systems under investigation can be summarized very simple in the Table 12.1, as follows:

The limited number of local cases is so taken to be reasonably representative of a wider category of municipal systems (Peters 1998). The cases are described in the real-life context in which they occurred (Yin 2003; Baxter and Jack 2008). Unsurprisingly, the boundaries between the individual phenomenon and its context are fluid. The analysis confronts the praxis of voter education with the four normative standards as described above.

Table 12.1 Institutional characteristics of cities

City	Vienna	Bregenz	Los Angeles
Formal autonomy	Medium	Low	High
Role of party politics	High	High	Low
Mayoral strength	High	High	High
Strength of direct democracy	Low	Medium	High

12.3.2 Poor Voter Information: The City of Vienna (Austria)

Overall, each Austrian municipality is the lowest level of government, under the supremacy of the federal and state governments. The autonomy of a municipality is relatively very low, as compared to some other federalist systems. The local organization, function, and activity are regulated by superior law. The mayor enjoys a strong formal and political position. The citizens are allowed to exercise certain rights of the people, so the informal participation has gained importance. The degree and quality of local direct democracy vary widely across the federal states of Austria, depending on the legislation of the federal states, and reaching from initiatives, referenda to citizens' assemblies (Eberhard 2013; Pleschberger 2003).

Vienna is the capital city of Austria. Vienna enjoys a singular formal position among all municipalities, as being simultaneously a federal state (Land), a municipality, and a statutory city. The local government system demonstrates striking political stability. The city council has been dominated by the Social Democratic Party (SPÖ) with either an absolute or a relative majority since the year 1920. The SPÖ is a strong member party, which brought it relatively secure victories in the elections in the past since 1920. Since the early 1990s, however, their margin of victory in the elections has been decreasing in the local elections on average, but nevertheless the party holds a very strong position. The mayor always comes from their ranks and has been elected indirectly, and serves a 5-year term. The mayor is the head of the executive branch and has some kinds of legislative power. In the executive sphere, the mayor is the "boss" of the city government. He is responsible for the implementation of the decisions of the council and the city government and can use the administrative resources for this purpose. His actions are bolstered by his strong party organization which he chairs normally.

The popular rights allow citizens to express and bring their political preferences into the local political process. Relatively late in the Austrian context, at the end of the 1970s, the development of popular rights spreads to the city constitution. During the reform debate representatives of the traditional ruling, SPÖ articulated their clear preference for representative democracy. They expressed strong skepticism about the role of direct democracy (Pleschberger and Mertens 2012). The professional representatives of the party till today tend to doubt the competence of citizens to decide on complex questions of urban development or to discredit instruments of direct democracy as a mere opinion survey. Direct democracy under these circumstances may often occur as a process structured and manipulated to legitimate fixed party decisions and not as a bottom-up process required and implemented by local citizens to decide on an urban issue according to their political beliefs.

The system of the citizens' rights allows citizens to express and bring their political preferences into the local political process. Since 1980, the city constitution has recognized the democratic rights of a binding and non-binding referendum. Eligible citizens have some power to recommend a novel policy to the local council, which the council can accept or ignore. They have no power to initiate and enact a specific policy without the approval of the local councilors. The council can initiate a consultative referendum by a simple majority, and at least five percent of registered voters must sign a petition to initiate a referendum. The right to initiate a binding referendum is the sole right of the municipal council. The transaction costs of the municipal council for the instrumentalization of the popular rights are low because in the political reality the SPÖ had the majority to initiate it.

Between 1973 and 2013, eight consultative referenda were held in Vienna, with a total of 25 questions posed. Two issues put to the vote of local citizens came from a citizens' petition, and two questions came from a local party. The vast majority of the questions puts mayors on their own in consultation with their ruling party, with other local council parties mostly left out of the process.

The municipal government simply has to inform the voters by sending a voting card which according to the city constitution provides to the voters "necessary information" about the voting process, the term has not been more clarified. More official information to educate the voter is not required to meet legal provisions.

- A voter who would like to support a petition has to bear his signature on a standardized form, which only contains in the masthead: the short name (title) and an abbreviated designation of the measure, both provided by the proponents of a petition. Further criteria such as length or neutrality of the formulation, and so on, do not have to be complied with. This gives opportunity on the part of proponents to use political tactics to raise a measure over the signature hurdle. Due to the small space available, the official form is not envisaged to facilitate an informative description of the measure.
- The official ballot card informs the voters about logistical details and the text of the voting question and that voter should answer yes or no to it.

Only two of all ballot questions contained financial notes (Pleschberger 2015). One concerned the extension of the services of the city subway system during the weekend, which the local council wanted to consult the people on, in 2010. The cost calculation was never made transparent in the released content despite urgent appeals in the council. At the same time four further measures were put to the vote on voting day, but without any financial notes, even though they had clear financial implications. The foregoing procedure was probably due to attempts to derail the measure at the ballot box, but the voters saw it differently, and voted for the extension of the subway service.

Another and instructive example of fiscal "informing" of the voter was the copyediting of a ballot question (previously used as a petition question), which was presented to the voters in December 1981. The measure was promoted by a city-wide political party. The background was a fierce dispute with the municipal government on the establishment of a new conference center on the periphery of the city. The ballot question claimed this would have been a "Milliardengrab" (a billions euro money pit). The voters followed the direction given to them by the proponent and rejected the measure by a large majority.

12.3.3 Voter Information at the Start: The City of Bregenz (Austria)

The City of Bregenz is the largest city and capital of Vorarlberg. The country state is often self-portrayed as a distinguished model for direct democracy in Austria by a vague association in popular memory with Swiss standards of direct democracy (Pleschberger 2015). That would mean a low influence of party politics and a high power of local voters to influence the local policy agenda. The mayor is elected directly and comes from different party ranks. He serves a 5-year term and his actions are bolstered by his strong party organization and he can at large rely on the permanent political support of his party. Since 1965, the citizenry has had three essential popular rights: Volksbefragung (Consultative Referendum), Volksbegehren (Petition Referendum), and Volksabstimmung (Binding Referendum). The binding referendum arises when the local council rejects the outcome of a petition or an initiative successfully jumps over the signature threshold. The mayor may also initiate a referendum. Binding referendums happened in 34 local councils, in which 38 issues have been presented to the voters (as of November 2015). They were initiated mostly by the local councils, and only a very few go back to starting as citizen initiatives.

Proponents of a desired measure draft the title and voting question which must be used after positive legal examination of the admissibility by the electoral authority on the petition form.

Since 2015, if a petition has qualified for the ballot, the mayor must inform the voters with a "vote brochure" which will be sent to them (it was formerly the "accompanying report"). The new brochure incorporates the following sections: logistical details of the direct vote, and the voting question; and as a deliberative element, the grounds of the measure in the view of the sponsors and of the members of the community board, which is to be displayed "as objectively as possible and if possible to the same extent." The grounds provided may be subsequently improved. The regulation introduces a very limited deliberative element because no argumentation can be adduced for or against by a third party. It may be the case that there is only positive or negative argumentation. It can also be questioned whether the mayor can guarantee an impartial procedure and the objectivity of the arguments presented. The mayor is a member of the local council and the local board, both organs are taking political positions on an issue. He is partisan and shall serve as the guard of objectivity or neutrality of voter information, the length of the argumentation, and so on. The presentation of the fiscal implications is not explicitly required, but it is not forbidden and maybe present by the responsible mayor.

The ballot card informs the voter on the voting question and repeats the text documented in the vote brochure.

An early example of the former praxis of voter information with the accompanying report can be found in the context of the referendum on the new urban development strategy in Bregenz. In the city since 1965, one consultative and two binding referenda were hold; the latter went back to a council decision. The local government had sent to the voters an accompanying report of four sheets, including the title page section, formal notice of the referendum, two favorable decisions of the whole council and the city board (Stadtrat), and the passing of the vote question. Attached was a planning sketch of the new concept. A fiscal hint indicated that the implementation of a positive decision was bound by financial possibilities. The deliberative element appeared in a marginal way. Furthermore, the newly introduced vote brochure is a misnomer. A recent practical example of such a vote brochure drafted in the small rural commune Reutte from September 2015 consists of five pages, including the title page. It contains the formal notice of the vote, opinions of the members of the local board, at different lengths. The vote question consists of eight words, and then comes two lines on the direct consequences of a yes or no outcome on the measure. The last page is a planning sketch. The deliberative element is reduced to the perspectives of the three members of the local board, two of which were in favor of the measure and one with an argument against it. The view of the citizens' initiative, which actually forced the binding referendum and rejected the designation of a new business location, did not appear in the vote brochure, what is legally permissible under the mayoral view.[2] The most recent example, a vote brochure from November 2015, drafted in the small urbanized commune of Lochau informs citizens to dispute over whether the town hall is to be renovated or to be rebuilt. The local council initiated the direct vote unanimously. The brochure consists of few pages and was similarly structured to the previous example. Interestingly, it contains no deliberative elements that represent the subject, with a pro or con perspective. Notes on fiscal implications are missing completely.

12.3.4 Voter Information at Maturity: The City of Los Angeles (California)

Los Angeles is an US charter law community, with a high autonomy to design its own form of government, allowing it to develop political and governmental autonomy. In 1999, the voters approved a new city charter that provides for a mayor–council type of local government system, and the directly elected mayor being the executive branch and the council the legislative. The mayor serves a 4-year term. As the executive officer, he is held responsible for the conduct of all city affairs and submits proposals and recommendations to the council. The mayor makes key appointments, presents the budget, and can veto legislation. In the current charter, the mayor is described as having "management authority" over the city government as well as being the chief executive officer of the city (Sonenshein

[2]Telephone conversation with the mayoress of Reutte on Wednesday, December 2, 2015 with the author.

2006, 42). But he does not rely in the real political sphere on permanent support of a local party.

Like the federal state of California, the City of Los Angeles is often described as a trend setting area of local democracy, what is unique in the international system, and comparable to the Swiss system. But in sharp contrast, the citizens are, according to a statement in the representative publication authored by Sonenshein (2006), uninterested in their local government and the official participation in the many elections has been relatively low.

California deserves a more detailed consideration. It is a hybrid democratic system (Baldassare and Katz 2008; Garrett 2005). The voters have the right to enforce and initiate legislation (Dubois and Feeney 1998; Eule 1990: 1509; Specht 2006; Kesselman 2011). The state at the national and at the local level (Gordon 2004, 2011; Graves 2012; Sonenshein 2006) is a trendsetting area, along with Oregon it vies for first place among the initiative states. Today an initiative can be submitted in every city and in every county of California (Gordon 2004). Nationwide over half of all cities have the initiative process, and all have a version of a referendum. The initiative process allows citizens to propose an entirely new measure and to get a ballot measure before the populace. That goes into effect if approved by a vote of the electorate at large. They can be statutory or amend the local charter. By the mechanism of referendum, the citizens can ratify or reject a statute or constitutional amendment. The empirical picture of local ballot measures is more complicated as at the state level.[3] Between 1990 and 2000, over 730 local initiative petitions circulated for signatures in the cities and counties (Gordon 2004). On average, just one initiative petition that circulated in a city, of a total of 482 local governments. The initiative activity was concentrated in just a few jurisdictions (predominantly San Francisco). From the local initiatives which qualified 45 % have been approved, little more than statewide (Public Policy Institute of California 2004). Of the citizen initiated measures related to boundaries of urban sprawl, urban development projects, term limits of elected officials, rent control, taxes for transportation or public safety, and reduction or repeal of utility user taxes (Gordon 2004; Levinson and Stern 2010: 695; Public Policy Institute of California 2004). And as the California Commission of Campaign Financing (1992: 143) has stated: “By far, the greatest share of countermeasures at the local level is referred measures placed on the ballot by the local governing body. It is very common for city councils or county boards to challenge an initiative by placing their own alternative measure on the same ballot. Local governments in Greater Los Angeles have demonstrated a strong propensity to react defensively when confronted with an initiative. Rarely have initiatives been adopted by the governing board without a vote of the people. More often, initiatives are rebuked by the body

[3]Municipal legislative bodies are required to file a report with the Secretary of State containing information regarding citizen-generated initiative measures, http://elections.cdn.sos.ca.gov//ballot-measures/pdf/county-initiative-2013-2014.pdf.

and passed onto the ballot along with an alternative legislative measure from which voters may choose."

The City of Los Angeles already introduced the ballot initiative in 1903, which gave citizens the opportunity to write important local measures and enforce them (California Commission of Campaign Financing 1992). The normal scenario of an initiative process is simple: a civil proponent drafts a measure, makes their intention known to the city, and municipal bodies assist with the design. After getting the official permission, proponents start a petition for collecting the necessary signatures, which varies according to the type of desired measure, and if successful, the city council will take up the measure or place it on the ballot for the next voting time (League of California Cities 2015).

The City of Los Angeles has a wide latitude in setting the detailed requirements of voter information (City of Los Angeles 2013). However, most local jurisdictions —general law as well as charter law communities—tend to follow the procedures dictated in the California Elections Code (Levinson and Stern 2010: 704–707; Elmendorf and Spencer 2013: 514). The city has the full responsibility for voter information (California Commission on Campaign Financing 1992: 192), which is closely linked to the different phases of the initiative process, and wherein multiple interactions take place in the course, where urban officials and the initiative proponents come together (League of California Cities 2015).

According to the Election Code of the City of Los Angeles, the voter information has to consider legal terms such as "impartial," "unbiased," "clear," "concise," "accurate," "readable," and so on, which are not precisely defined and sound very promising in the normative sense (City of Los Angeles 2015).

Proponents of a measure submit to the city clerk a "notice of intention" for the circulation of a petition containing a statement about, and the title of their proposed measure. If there is no rejection on formal grounds, the official sends a request to the city attorney to prepare an official petition title and a fair and impartial official petition summary of the primary provisions of the proposed ordinance to be included on all copies of the petition form to be circulated by the proponents for signatures. This document should simply summarize, in 500 words or less, what the measure says. It shall be written in language as not to be intentionally an argument or designed to create prejudice either for or against the measure and shall not be false or misleading. The official title must be printed across the top of each page of the petition. The official description is used in subsequent notifications of intent to circulate in newspaper publications (California Commission on Campaign Financing 1992: 4–5).

The voter information pamphlet (City of Los Angeles 2015 secs. 402–408) consists of the following sections, at the top the ballot summary, following ballot title, the question to be submitted to the voters, the impartial summary of a measure, the financial impact statement and/or tax rate statement, arguments and rebuttal arguments for and against, the official title and summary prepared by the city attorney, the whole text of the measure is at the end of the pamphlet.

The ballot summary (sec 404 subsec c) contains six subsections: title, the issue, the situation, the proposal, a yes vote means, and a no vote means. The maximum word limit is 300 (the title is excluded) and can be increased in conjunction with the will of the President of the council. The special Ballot Simplification Committee drafts a simplified digest city of the measure. It may at its own discretion, utilize readability formulas and so on, which the committee considers appropriate and effective in achieving the closest proximity to the eighth-grade level of readability. Commonly used is the Flesch–Kincaid formula, that uses the number of syllables per 100 words and the average number of words per sentence to assess the readability of the text (Center for Governmental Studies 2008: 234). The simplified version of the ballot measure shall have the form of a fair and impartial ballot summary, with the overall goal, that the average voter can understand the summary (subsecs. a, b). Simplified digests are subject to approval by the city council, the council has no authority to amend or change digests; it can only approve or reject them as written. If the council rejects the digest then the pamphlet is assembled without any simplified digest for that measure.

The impartial summary prepared by the Chief Legislative Analyst consists of a maximum of 600 words (sec. 604 subsec. a), which can be increased under certain reasons. While the general law jurisdictions only have to provide a simple fiscal note, Los Angeles must provide by the city administration officer a financial impact statement with a maximum of 150 words, which can be increased (for reasons of the type of measure, complexity, length). It gives "an estimate of the amount of any increase or decrease in revenues or costs to city government, or an opinion as to whether or not a substantial net change in city finances would result …" (sec. 604 subsec. b). Increased costs "shall be set forth in boldface print in the Voter Information Pamphlet" (id.). Further the analysis "shall be written in clear, concise terms which will be easily understood by the average voter and wherever possible shall avoid the use of technical terms" (id.). The financial analysis of a measure has to appear in the petition, voter information pamphlet, and ballot question (California Commission on Campaign Financing 1992, 312; IRI 2000, 8; National Conference of State Legislatures 2002, 27).

Arguments and rebuttal arguments for and against the measure, if any, are to be incorporated in the voter pamphlet (sec. 402 subsec. c). They represent the deliberative section of voter information and they can provide insights that more neutral attorney general or legislative analyst summaries may avoid (Center for Governmental Studies 2008: 235). The argument and rebuttal argument shall be written by the proponents (Election Code of the City of Los Angeles, sec. 605 subsec. b). The length of the arguments for or against a measure shall not be more than 1000 words. Each argument shall not exceed 300 words for or against any measure sec. 605 subsec. e). All arguments shall be signed by the authors. The invited persons or organizations must ensure that their arguments, by their own conscience are true and correct, as far as these lie within their powers. They should have no personal interest in the measure to be considered. The arguments for or against a measure can have a word total of 300, and rebuttal argument shall not exceed 150 words each.

The official ballot title is to be a "short, impartial statement in bold capital letters indicating generally the stated purpose of the measure" (sec. 602 subsec. b). In addition the ballot question shall be presented in the form of a question and consist of an "impartial statement of the purpose of the measure." The ballot title and question combined shall not exceed 175 words, except as to bond propositions or as otherwise limited by state law in consolidated elections (sec. 602 subsec. c). Both actions shall not alter the meaning of the official petition title and official summary petition.

The whole text of the measure shall include the provisions of the proposed measure and the existing provisions of law repealed or revised by the measure. The differing from the existing provisions of law shall be distinguished in print, so that the effect of the proposed measure upon existing provisions can be determined (sec. 402 subsec. c).

12.3.5 Practical Example: What Gets the Voter About the Medical Marijuana Propositions (2013)

Repeated debates and reports on the spread of illegal dispensaries for marijuana for medical purposes formed the background for two popular initiatives and a counter initiative in 2013. Thus, the official voter information pamphlet about the "Medical Marijuana Regulation and Taxation Ordinance" described three proposals (Measures D, E, F).[4] With proposition D, which prevailed at the ballot box on May 21, 2013, the council reacted to the popular initiatives. The propositions are described in the voter pamphlet in the same structure. Measure D was made out over 21 pages and structured as follows: a full legal copyedit of the measure (10), arguments for and against the measure (7), a popular summary (1), an impartial summary (1), the voting question (0.5), and a financial impact statement (0.3).

The popular ballot summary focused on few aspects in simplistic words and sounds redundant.

The voter question, beginning with "shall …" and listed sequentially in extreme density, in turn the important aspects about the measure.

The impartial analysis described the important aspects of the measure in more detail than the text of the voter question.

The financial impact statement deepened the info about the financial implications and suggests, that the measure will reduce the number of Medical Marijuana Businesses (MMBs), which gave a vaguely associated causality, with the increase in business tax. The exact impact of the measure on the public revenue could not be quantified (it cited the take of the 2012 financial year). Any possible revenue loss may be offset by the tax increase. Following was indicated that the revenue from the business tax would flow into the general fund of the city, from which the public

---

[4]http://ballotpedia.org/City_of_Los_Angeles,_California_elections.

services are financed (police, fire, and so on). Whether it would lead to higher spending on health care or enforcement expenditures, was described as unknown. The Financial Impact Statement had the briefest scope of all sections.

In the “pro” and “con” section of the pamphlet 14 signatories presented their arguments for or against the measure. The majority of them can be associated with major economic, political or organized interests. A proponent of the measure D refers to himself as an AIDS patient.

The full text of the measure describes the measure as a legislative piece, which will be directly applicable in the case it is approved.[5]

12.3.6 *Practical Standards as Issues for Debate and Reforms*

A good voter pamphlet and other information sources serve voters to become informed. The listed practical standards are key elements to reach this objective, and further more would make sense (e.g., visual appeal, font size). Statewide surveys indicate voter guides give useful information (Brien 2002; Silva 2000: 24–27). Voters get the pamphlet for free, take it, and read it carefully. The data aggregates the self-reporting of voters, who naturally have a tendency to present themselves as good citizens (Elmendorf and Spencer 2013: 519). According to pragmatic diagnoses of scientific observers, the voters did not take the time in everyday life to read substantial parts of the voter information; such as the ballot argument, summaries nor the full language of a proposed initiative (Schacter 1995; Sutro 1994: 954).

It can be found among the many examples of voter pamphlets (Dubois and Feeney 1998: 169–175) and with local examples too, that those with more than 1000 pages in length give a significant warning of information overload (Garrett and McCubbins 2008: 55–57). Professor Eule, a law specialist, received in 1988 a state ballot pamphlet with 159 pages, he described it as like an unreadable phonebook. A short time later followed a ballot pamphlet from the City of Los Angeles with 74 pages, it was in slightly larger font (Eule 1990, 1991). A quantitative excess of information can cause confusion, disinterest, and promotes informational ignorance that does not stop even before legally qualified voters (Eule 1990: 1509). The long and complex language of official voter guides are plausible good reasons for the voters deliberately ignoring the information. The length is often too long, and it is estimated that the average voter would have taken a double digit numbers of hours to read the pamphlet (Sutro 1994). It is a well-established plausible claim that many voter pamphlets (not all) are easily too long and

[5]The proposition D became e ffective before the date of publication, that is upon June 20, 2013 with Ordinance 182580 Enacting Prop D; http://www.lacityattorney.org/#!medical-marijuana/cgh5
.

disrespect the temporal and intellectual capacities of many voters (Eule 1990, 1991; California Commission on Campaign Financing 1992: 2; Center for Governmental Studies 2008: 9; Sutro 1994: 951–954; Reilly and Walker 2010: 227; Kruse 2001: 145–146; Schacter 1995: 142). Significant shorter official sources would better contribute to informing the voter (Ford and Kemokai 2014: 209–210). A maximum word limit of 5000 words is proposed for the description of a ballot measure (California Commission on Campaign Financing 1992: 3). With the expectation that it "would encourage proponents to limit their causes and describe them in simpler language" (California Commission on Campaign Financing 1992: 15). The tricky question of what is the correct length for a voter-friendly pamphlet cannot be answered of course in general.

Political blockages in the representative system stimulate flight into direct democracy and the growth of ballot initiatives (California Commission on Campaign Financing 1992: 6–7). Behind each proposition stands political and economic forces that promote, prevent, or want to change something. On a voting day it may be there are many and long measures to consider (Kruse 2001: 145–146). Counter initiatives would like to destroy an initiative and are thus launched on tactical grounds (California Commission on Campaign Financing 1992: 2). If a proposition clearly follows the single-subject rule-(Sutro 1994) it is authorized. To solve the mere quantity problem of organized direct democracy, the number of initiatives admitted to a voting day should be limited (California Commission on Campaign Financing 1992: 154). This brings with it the danger that the well-sponsored initiatives are favored and can qualify for the vote fastest.

Even one of the most serious criticism of the initiative process seems to be the assertion, that voters do not always fully understand the contents of the initiatives on which they are asked to vote, because the drafting makes them too complex (National Conference of State Legislatures 2002, 44). This is confirmed by opinion polls (Center for Governmental Studies 2008: 90; Silva 2000: 15). Currently, 78 % of adults and 83 % of likely national voters in California say "that initiative wording is often too complicated, making it confusing to understand what would happen if an initiative passed" (Public Policy Institute of California 2013: 5). Official title and summary are not simple enough for many voters (Silva 2000: 12). According to a widely cited study on statewide ballots, the understanding of the title and summary requires a readability level which is far above the average of the population (Magleby 1984: 118, as quoted e.g., from Kruse 2001: 144–145). These assertions are repeatedly circulated (e.g., Schacter 1995: 139–140). Some observers, such as Eule wonder polemically "that anyone knows what he or she is voting on" (Eule 1990: 1516), even when an educated voter marks the space on a paper (Eule 1991: 786). You do not need a readability test to make the claim that the description of the full text of a measure in legal terms is incomprehensible for the majority of voters. As Schacter (1995: 128) states, in particular voters lack any "detailed knowledge of the legal context surrounding a proposed initiative," so it is highly predictable they do not know the technical language. The California Political Reform Act 1974 has in retrospect not improved the comprehensibility of texts (Center for Governmental Studies 2008: 48). The demand, that initiative measures should be written for the average educated voter still remains an undisputed general recommendation in the official voter information debate. As Schacter (1995: 141) critically states: "This is an ironic state of affairs, for the widely acknowledged impenetrability of statutory language is a principal reason that many States require the government to provide more readily understandable explanations of ballot measures to voters."

Finally, the practical implementation of voter education is structured as an interactive process between proponents and appointed or elected administrative officials and political representatives who enable the politicization of voter education. Accordingly, official information sources should be drafted in an objective, neutral, and impartial manner. Public officials can use their position and roles assigned to influence the direct democratic process according to their political motives (Burnett and Kogan 2015: 110). As Garrett (2001: 171) states: "Even if the government attempts to stay neutral in its explanation of the referendum and the underlying issues and saves its advocacy for another portion of the pamphlet, the appearance of bias is hard to avoid." And as the California Commission on Campaign Financing (1992: 189) suggested long time ago, in an often quoted sentence, "Even assuming a competent and honest officeholder, a certain amount of bias in the drafting of the descriptions may inevitably emerge from the individual drafter's own values and experiences." The California Commission on Campaign Financing (1992: 9) suggested the publication of the rules and standards used or the holding of public hearings in the early initiative process to reduce the bias in the formulation of an initiative. Responsible public officials meet proponents, during the initiative process, and the latter want to help their initiative to succeed and to quickly jump the signature hurdles. They consider the petition title as a critical success factor and offer some kind of informal assistance with drafting the official title. Accordingly, they are influenced by their interests (Kim 2010; California Commission on Campaign Financing 1992). Proponents tend to draft "their" title in with a vague and positive language. Whereas the attorney, an elected official, drafts the caption and summary of a petition, that normally becomes the official description of the initiative in the voters' pamphlet and on the ballot (California Commission on Campaign Financing 1992: 186). The official version will seek a descriptive, neutral language, which makes the text more sober and less attractive a political title (Kim 2010; Hastings and Cann 2014: 118). If a titling is contrary to the interests of the proponents, they have the possibility to compensate in practice by the way that they go about the signature collection, with misleading slogans and emotional appeals (Smit 2007: 265).

The fiscal information about a measure is another possible area of political interests and possible bias. It has to appear on the petition, in the voter pamphlet, and on the ballot question in different descriptions. For instance, the goal to stabilize the municipal budget against a spending excess may conflict with the right of initiatives to force the local government to spend money for new public services (Levinson and Stern 2010: 695). A city administrative officer responsible for drafting the financial impact statement may dramatize the financial effects of a proposed measure. In fairness, there are many cost effects of a measure which are "unknowns," which can cause discomfort to fiscally conservative voters and perhaps motivate them toward a no vote. Or by simply referring or to emphasizing the overall budgetary context in terms of a local fiscal crises. This seems trustworthy, as local governments possess superior information and expertise about the fiscal situation in the municipality.

Another space for political interests is the section with pro or con arguments in the voter guide. This is very likely, as voters will tend to consult this section, as with the title and summary, of a measure in their own way (Primo 2013, 116). This section presents the most recognized arguments from the most organized interests. This does not automatically mean that the best arguments are presented, but often probably the strongest and loudest interests put their views.

12.4 Conclusions

Although the use of local direct democratic decisions on propositions is quite common, we can see that three local governments in two national contexts regulate and implement the task of voter information in divergent ways, which has potentially important but different impacts on urban governance processes. In the City of Vienna any kind of substantial official voter information is non-existent, in the absence of a set of guiding legal provisions. The approach in the City of Bregenz demonstrates some moderate signs to proceed in the right direction. Whereas, in the mature case, the City of Los Angeles operates within the framework of a detailed set of statutory regulations that refer to ambitious normative and procedural provisions that shall guide the official voter information in practice.

One result that has emerged clearly is that history and institutions matter. The regulatory difference of voter information is embedded in the divergent history and the broader institutional context of local government systems. A long-enduring history and strong institutionalization of local direct democracy might contribute to a good official voter information. This assertion does not assume that history leads to an efficient equilibrium, and we know many possibilities of inefficiencies (March and Olsen 2008). If an agency performance is perceived as defect or weak, a strong mayor would have the power to focus his attention productively on any existing procedure and could potentially pursue a policy of change in the framework of the legal and political constraints. In the realm of official voter information, it further matters whether the mayor has a tight knot with the local party systems. The politico-institutional overlapping of the mayoral role with the interests of a local party seemingly is a crucial constraint. Local party politics tends to distrust direct democracy. The ordinary citizens are not reasonable voters, they cannot be trusted, and why inform them according to the highest normative standards makeable if a political system generally is better off with representative lawmaking.

First, the politicization of voter information is ever present, certainly very different. In the Viennese case, the party politics and public officials enjoy a very free hand to manipulate the information task in their own political interests without any legal guidance or restrictions. In the Bregenz case the mayor, as an elected party politician, has to concern himself with a few vague standards and draft and provide a vote brochure. What we can learn from the mature case of the City of Los Angeles is the fact that the very detailed regulation of the voter information leads to the politicization of details of voter information, prompting questions that sometimes seem simply practical by nature, but are deeply rooted in the local political arena.

Second, the more ambitious the standards of voter information, the higher are the bar a local government has to achieve. These management issues are basically of non-controversial nature, but are still out of reach of a concrete answer in the more technical sense. This undertaking would have to concern the constant claims provided by researchers to design the information for the average voter. It sounds right, but still a lot of questions to be answered in a more technical sense. What is the appropriate length of a voter pamphlet? Is the information outlet accessible and readable to the majority of voters, and is it drafted without any political bias? As the vivid reform debate demonstrates, it remains open whether these questions ever can be answered definitely by interested researchers and/or innovative local practitioners. The list of disputed questions is far longer than the list of operational reform proposals put into practice (e.g., Baker 1991: 772–775; Silva 2000; Miller 2001: 1073; Schrag 2001; Burnett et al. 2010: 317–324). At the minimum, the official power to inform should provide the voter with useful information as much as possible about the facts to better understand the "points of sale"—and give additional information if necessary. The voter shall be provided with information about the measure and the essential fiscal implications. What are the direct costs of the proposed measure and the effects on local revenue? Such pragmatic claims should be non-controversial and broadly acceptable in different institutional contexts. But as we know, they are not like other institutional innovations. Taking into account seriously it refers to the challenge of if necessary regulating of local government, that citizens would be better off collectively if they all became better informed on a measure to decide by them (Clark and Lee 2003).

Where it is possible and necessary the national government could interfere and use its coercive powers to regulate the voter information. But local governments would be better off when they assimilate new information voluntarily. The other view, arising from comparative government studies, is that local government officials are self-interested in getting a deeper understanding of other cases, learn from the actions of other localities around the globe, and as a follow-up becoming improving their own practice by introducing innovations (Elcock 2008). They can learn from other advanced localities in their own country or from foreign systems. This may lead to a slow convergence of understandings and practices, which in turn might result in a common way of approaching voter information in a direct local democracy. Comparative studies can contribute to ease the flow of information in this direction. Contrarily, the realities under study give no grounds to hope that lessons drawing will take place easily. In the author's view, the City of Vienna already could have learnt a little from the localities in the Austrian federal state of Vorarlberg, if Vienna would have been interested to experiment with the vote brochure as a new and innovative tool. This policy alone would improve matters considerably. But the city still lacks any attempts to discuss and to reform the law and practice of official voter information. It is a non-decision yet. Both Austrian localities under investigation could take an ambitious look abroad and learn from a mature case, because a different politics of voter information is visible and viable, according to the many and diverging information sources. It is at least theoretically acknowledged that the imitation of larger cities is a common pattern of policy diffusion. But in the course of reality, this wisdom not always may work as it is theoretically possible and expected.

References

Baldassare M, Katz C (2008) The coming of age of direct democracy: California's recall and beyond. Public Opin Q 72:394–397
Baker LA (1991) Direct democracy and discrimination. A public choice perspective. Chic Kent Law Rev 67:707–776
Baxter P, Jack S (2008) Qualitative case study methodology: study design and implementation for novice researchers. Qual Rep 13:544–559
Bendor J, Bullock JG (2008) Lethal incompetence: voters, officials and systems. Crit Rev 20:1–23
Benz M, Stutzer A (2004) Are voters better informed when they have a larger say in politics? Evidence for the European Union and Switzerland. Public Choice 119:31–59
Berelson PR, Lazarsfeld PF, McPhee WN (1954) Voting. A study of opinion formation in a presidential Campaign. University of Chicago Press, Chicago
Brien P (2002) Voter pamphlets: the next best step in election reform. J Legis 28:87–112
Buchanan JM (1954) Individual choice in voting and the market. J Polit Econ 62:334–343
Bullock JG (2011) Elite influence on public opinion in an informed electorate. Am Polit Sci Rev 105:496–515
Burnett C, Garrett E, McCubbins MD (2010) The dilemma of direct democracy. Elect Law J 9:305–324
Burnett CM, McCubbins MD (2013) When common wisdom is neither common nor wisdom: exploring voters' limited use of endorsements on three ballot measures. Minn Law Rev 57:1557–1595
Burnett CM, Kogan V (2015) When does ballot language influence voter choices? Evidence from a survey experiment. Polit Commun 32:109–126
California Commission on Campaign Financing (1992) To govern ourselves. Ballot initiatives in the Los Angeles area. Center for Governmental Studies, Los Angeles
Carter ME (2011) Regulating abortion through direct democracy: the liberty of all versus the moral code of a majority. Boston Univ Law Rev 91:305–346
Center for Governmental Studies (2008) Democracy by initiative: shaping California's Fourth Branch of Government. Center for Governmental Studies, Los Angeles
City of Los Angeles (2013) Initiative, referendum & recall petition handbook. Los Angeles. http://clerk.lacity.org. Accessed 25 Dec 2015
City of Los Angeles (2015) Election code of the city of Los Angeles. Los Angeles http://cityclerk.lacity.org. Accessed 25 Dec 2015
Clark JR, Lee DR (2003) Regulating government. In: Rowley CK, Schneider F (eds) The encylopedia of public choice. Kluwer, Norwell, pp 482–484
Cronin T (1989) Direct democracy: the politics of initiative, referendum, and recall. Harvard University Press, Cambridge MA
Delli Carpini MX, Keeter S (1996) What Americans know about politics and why it matters. Yale University Press, New Haven, CT

Downs A (1957) An economic theory of democracy. Harper & Brothers, New York
Dubois PL, Feeney F (1998) Lawmaking by initiative: issues, options, and comparisons. Agathon Press, New York
Eberhard H (2013) Austria—municipalities as the “third tier” of Austrian Federalism. In: Panara C, Varney M (eds) Local Government in Europe. The ‘fourth level’ in the EU multilayered system of governance. Routledge, New York, pp 1–25
Elcock H (2008) Elected mayors: lesson drawing from four countries. Public Adm 86:795–811
Elmendorf CS, Spencer DM (2013) Are ballot titles biased? Partisanship and ideology in California’s supervision of direct democracy. U.C. Irvine Law Rev 3:511–549
Elmendorf CS, Schleicher D (2013) Informing consent: voter ignorance, political parties, and election law. Univ Ill Law Rev 2013:363–432
Eule J (1990) Judicial review of direct democracy. Yale Law J 99:1503–1590
Eule J (1991) Representative government: the people’s choice. Chic Kent Law Rev 67:777–790
Fiorina MP (1990) Information and rationality in elections. In: Ferejohn J, Kuklinski J (eds) Information and democratic processes. University of Illinois Press, Urbana, pp 329–342
Ford P, Kemokai T (2014) Direct democracy: a global comparative study on electoral initiative and referendum mechanisms. http://www.mcgeorge.edu/Documents/Publications/directDemocracyNov2014.pdf. Accessed 24 Dec 2015
Funk P, Gathmann C (2014) Direct democracy as a disciplinary device on excessive public spending. CESifo DICE Rep 12:18–23
Gafke R, Leuthold D (1979) The effect on voters of misleading, confusing, and difficult ballot titles. Public Opin Q 43:394–401
Garrett E (2001) Issues in implementing referendums in Israel: a comparative study in direct democracy. Chic J Int Law 2:159–182
Garrett E (2005) Hybrid democracy. George Wash Law Rev 73:1096–1274
Garrett E, McCubbins MD (2008) When voters make laws—how direct democracy is shaping American cities. Public Works Manag Policy 13:39–61
Gordon TM (2011). Initiatives aren’t as bad as you think. Brookings Opinion October 4. http://www.brookings.edu/research/opinions/2011/10/04-ballot-initiatives-gordon. Accessed 24 Dec 2015
Gordon TM (2004) The local initiative in California. Public Policy Institute of California, San Francisco
Graves L (2012) Local ballot initiatives. The Lucy Burns Institute, Middleton. http://www.lucyburns.org. Accessed 24 Nov 2015
Hastings J, Cann D (2014) Ballot titles and voter decision making on ballot questions. State Local Gov Rev 13:1–10
Jakubowski P, Tegner H, Kotte S (1997) Strategien umweltpolitischer Zielfindung: eine ökonomische Perspektive. Lit Verlag, Münster
Kesselman D (2011) Direct democracy on election day: ballot measures as measures of American democracy. Transatlantica 2011:1–13
Kim D (2010) Don’t judge an initiative by its title: a report on the titles of California ballot measures. www.mcgeorge.edu/documents/…/report3.pdf. Accessed 24 Nov 2015
Kirchgässner G, Feld LP, Savioz MR (1999) Die direkte Demokratie. Modern, erfolgreich, entwicklungs- und exportfähig. Verlag Franz Vahlen, München
Kruse B (2001) The truth in masquerade: regulating false ballot proposition ads through state anti-false speech statutes. Calif Law Rev 89:129–181
League of California Cities (2015) Initiatives/Referendums. https://www.cacities.org. Accessed 24 Nov 2015
Leib EJ, Elmendorf CS (2012) Why party democrats need popular democracy. Calif Law Rev 100:69–114
Levinson JA, Stern RM (2010) Ballot box budgeting in California: the bane of the golden state or an overstated problem? Hastings Const Law Q 101:689–744
Lupia A (1994) Shortcuts versus encyclopedias: information and voting behavior in California insurance reform elections. Am Polit Sci Rev 88:63–76

Lupia A, Matsusaka JG (2004) Direct democracy: new approaches to old questions. Annu Rev Polit Sci 7:463–482

Magleby DB (1984) Direct legislation: voting on ballot propositions in the United States. Johns Hopkins University Press, Baltimore

March JG, Olsen JP (2008) Elaborating the "new institutionalism". In: Rhodes RAW, Binder SA, Rockman BA (eds) The Oxford handbook of political institutions. Oxford University Press, Oxford, pp 3–20

Miller, Kenneth P (2001) Constraining populism: the real agenda of initiative reform. Santa Clara L Rev 41:1037–1084

National Conference of State Legislatures (2002) Initiative and referendum in the 21st century final report and recommendations of the NCSL I&R Task Force. http://www.ncsl.org/research/elections-and-campaigns/task-force-report.aspx. Accessed 24 Nov 2015

Peters BG (1998) Comparative politics: theory and methods. NYU Press, New York

Pleschberger W (2003) Cities and municipalities in the Austrian political system since the 1990s. New developments between "efficiency" and "democracy". In: Kersting N, Vetter A (eds) Reforming local government in Europe. Closing the gap between democracy and efficiency. Westdeutscher Verlag, Opladen, pp 113–136

Pleschberger W (2015) Kommunale und direkte Demokratie in Österreich—Strukturelle und prozedurale Probleme und Reformvorschläge. In: Öhlinger T, Poier K (eds) Direkte Demokratie und Parlamentarismus. Böhlau Verlag, Wien-Köln-Mainz, pp 359–395

Pleschberger W, Mertens Ch (2012) Zur Parteipolitisierung der direkten kommunalen Demokratie. Am Beispiel der Großstadt Wien. PRuF Mitteilungen 18:24–35

Pommerehne WW (1978) Institutional approaches to public expenditure: empirical evidence from Swiss municipalities. J Public Econ 9:255–280

Pommerehne WW, Schneider F (1978) Fiscal illusion, political institutions, and local public spending. Kyklos 31:381–408

Popkin S (1994) The reasoning voter: communication and persuasion in presidential campaigns, 2nd edn. University of Chicago Press, Chicago

Primo D (2013) Information at the margin: campaign finance disclosure laws, ballot issues, and voter knowledge. Elect Law J 12:114–129

Public Policy Institute of California (2004) How do Californians use local ballot initiatives? Research Brief 93. http://www.ppic.org/content/pubs/rb/RB_904TGRB.pdf. Accessed 24 Nov 2015

Public Policy Institute of California (2013) The California initiative process— How democratic is it? Occasional papers. http://www.iandrinstitute.org/docs/CA-Commission-How-Democratic-Is-It-IRI.pdf. Accessed 24 Nov 2015

Rapeli L (2014) The conception of citizen knowledge in democratic theory. Palgrave Macmillan, Basingstoke

Reilly S (2013) Design, meaning and choice in direct democracy. The influences of petitioners and voters. Ashgate, Aldershot

Reilly S, Richey S (2011) Ballot question readability and roll-off: the impact of language complexity. Polit Res Q 64:59–67

Reilly S, Walker C (2010) Judicial elections' impact on participation in direct democracy. Justice Syst J 31:225–241

Sartori G (1992) Demokratietheorie. Wissenschaftliche Buchgesellschaft, Darmstadt

Schacter JS (1995) The pursuit of "popular intent": interpretive dilemmas in direct democracy. Yale Law J 105:107–176

Schrag P (2001) Symposium, The fourth branch of government? You bet. Santa Clara L Rev 41:937–949

Silva JF (2000) The California initiative process: background and perspective. Public Policy Institute of California. http://www.dcfn.ppic.org/content/pubs/op/OP_1100FSOP.pdf. Accessed 24 Nov 2015

Smit G (2007) Solving the 'initiatory construction' puzzle (and improving direct democracy) by appropriate refocusing on sponsor intent. Univ Colo Law Rev 78:257–305

Somin I (2013) Democracy and political ignorance. Why smaller government is smarter. Stanford University Press, Redwood City

Sonenshein RJ (2006) Los Angeles. Structure of a city government. League of Women Voters of Los Angeles, Los Angeles

Specht A (2006) The government we deserve? Direct democracy, outraged majorities, and the decline of judicial independence. Univ St. Thomas Law J 4:132–156

Sutro SH (1994) Interpretation of initiatives by reference to similar statutes: canons of construction do not adequately measure voter intent. Santa Clara Law Rev 34:945–976

Tausanovitch Ch, Warshaw Ch (2014) Representation in municipal government. Am Polit Sci Rev 108:605–641

Theodore NR (2013) We the people: a needed reform of state initiative and referendum procedures. Mont Law Rev 78:1401–1449

Tolbert CJ, McNeal RS, Smith DA (2003) Enhancing civic engagement: the effect of direct democracy on political participation and knowledge. State Polit Policy Q 3:23–41

Tolbert CJ, Smith DA (2006) Representation and direct democracy in the United States. Representation and Journal of Representative Democracy 42:25–44

Warner DM (1995) Direct democracy: the right of the people to make fools of themselves. The use and abuse of initiative and referendum, a local government perspective. Seattle Univ Law Rev 19:47–100

Yin RK (2003) Case study research: design and methods, 3rd edn. Sage, Thousand Oaks

Index

© Springer International Publishing Switzerland 2017

C. Nunes Silva and J. Buček (eds.), *Local Government and Urban Governance in Europe*, The Urban Book Series, DOI 10.1007/978-3-319-43979-2

GPSR Compliance

The European Union's (EU) General Product Safety Regulation (GPSR) is a set of rules that requires consumer products to be safe and our obligations to ensure this.

If you have any concerns about our products, you can contact us on ProductSafety@springernature.com

In case Publisher is established outside the EU, the EU authorized representative is:

Springer Nature Customer Service Center GmbH
Europaplatz 3
69115 Heidelberg, Germany

Batch number: 10756376

Printed by Printforce, the Netherlands